ENGINES OF CREATION

K. ERIC DREXLER

Engines of Creation

Foreword by Marvin Minsky

ANCHOR BOOKS
DOUBLEDAY
NEW YORK LONDON TORONTO SYDNEY AUCKLAND

AN ANCHOR BOOK
PUBLISHED BY DOUBLEDAY
a division of Bantam Doubleday Dell Publishing Group, Inc.
666 Fifth Avenue, New York, New York 10103

ANCHOR BOOKS, DOUBLEDAY, and the portrayal of an anchor
are trademarks of Doubleday, a division of Bantam Doubleday
Dell Publishing Group, Inc.

Engines of Creation was originally published in hardcover
by Anchor Books in 1986, and in paperback in 1987.

Library of Congress Cataloging-in-Publication Data
Drexler, K. Eric.
    Engines of creation.
    Bibliographic notes: p.
    Includes index.
        1. High technology.   2. Twentieth century—
Forecasts.   I. Title
T47.D74   1986   600   85-25362
ISBN 0-385-19973-2 (pbk.)

FOREWORD

K. ERIC DREXLER'S *Engines of Creation* is an enormously original book about the consequences of new technologies. It is ambitious and imaginative and, best of all, the thinking is technically sound.

But how can anyone predict where science and technology will take us? Although many scientists and technologists have tried to do this, isn't it curious that the most successful attempts were those of science fiction writers like Jules Verne and H. G. Wells, Frederik Pohl, Robert Heinlein, Isaac Asimov, and Arthur C. Clarke? Granted, some of those writers knew a great deal about the science of their times. But perhaps the strongest source of their success was that they were equally concerned with the pressures and choices they imagined emerging from their societies. For, as Clarke himself has emphasized, it is virtually impossible to predict the details of future technologies for more than perhaps half a century ahead. For one thing, it is virtually impossible to predict in detail which alternatives will become technically feasible over any longer interval of time. Why? Simply because if one could see ahead *that* clearly, one could probably accomplish those things in much less time—given the will to do so. A second problem is that it is equally hard to guess the character of the social changes likely to intervene. Given such uncertainty, looking ahead is like building a very tall and slender tower of reasoning. And we all know that such constructions are untrustworthy.

How could one build a sounder case? First, the foundations must be very firm—and Drexler has built on the soundest areas of present-day technical knowledge. Next, one must support each important conclusion step in several different ways, before one starts the next. This is because no single reason can be robust enough to stand before so many unknowns. Accordingly, Drexler gives us multiple supports for each important argument. Finally, it is never entirely safe to trust one's own judgments in such matters, since all of us have wishes and fears which bias how we think—without our knowing it. But, unlike most iconoclasts, Drexler has for many years courageously and openly exposed these ideas to both the most conservative skeptics and the most wishful-thinking dreamers among serious scientific communities like the one around MIT. He has always listened carefully to what the others said, and sometimes changed his views accordingly.

Engines of Creation begins with the insight that what we can do depends on what we can build. This leads to a careful analysis of possible ways to stack atoms. Then Drexler asks, "What could we build with those atom-stacking mechanisms?" For one thing, we could manufacture assembly machines much smaller even than living cells, and make materials stronger and lighter than any available today. Hence, better spacecraft. Hence, tiny devices that can travel along capillaries to enter and repair living cells. Hence, the ability to heal disease, reverse the ravages of age, or make our bodies speedier or stronger than before. And we could make machines down to the size of viruses, machines that would work at speeds which none of us can yet appreciate. And then, once we learned how to do it, we would have the option of assembling these myriads of tiny parts into intelligent machines, perhaps based on the use of trillions of nanoscopic parallel-processing devices which make descriptions, compare them to recorded patterns, and then exploit the memories of all their previous experiments. Thus those new technologies could change not merely the materials and means we use to shape our physical environment, but also the activities we would then be able to pursue inside whichever kind of world we make.

Now, if we return to Arthur C. Clarke's problem of predicting more than fifty years ahead, we see that the topics Drexler treats make this seem almost moot. For once that atom-stacking process starts, then "only fifty years" could bring more change than all that had come about since near-medieval times. For, it seems to me, in spite of all we hear about modern technological revolutions, they really haven't made such large differences in our lives over the past half century. Did television really change our world? Surely less than radio did, and even less than the telephone did. What about airplanes? They merely reduced travel times from days to hours—whereas the railroad and automobile had already made a larger change by shortening those travel times from weeks to days! But *Engines of Creation* sets us on the threshold of genuinely significant changes; nanotechnology could have more effect on our material existence than those last two great inventions in that domain—the replacement of sticks and stones by metals and cements and the harnessing of electricity. Similarly, we can compare the possible effects of artificial intelligence on how we think—and on how we might come to think about ourselves—with only two earlier inventions: those of language and of writing.

We'll soon have to face some of these prospects and options. How should we proceed to deal with them? *Engines of Creation* explains how these new alternatives could be directed toward many of our most vital human concerns: toward wealth or poverty, health or sickness, peace or war. And Drexler offers no mere neutral catalog of possibilities, but a multitude of ideas and proposals for how one might start to evaluate them. *Engines of Creation* is the best attempt so far to prepare us to think of what we might become, should we persist in making new technologies.

MARVIN MINSKY
Donner Professor of Science
Massachusetts Institute of Technology

ACKNOWLEDGMENTS

THE IDEAS IN THIS BOOK have been shaped by many minds. All authors bear an incalculable debt to earlier writers and thinkers, and the Notes and References section provides a partial acknowledgment of my debt. But other people have had a more immediate influence by reading and criticizing all or part of the several papers, articles, and draft manuscripts ancestral to the present version of this book. Their contributions have ranged from brief letters to extensive, detailed criticisms, suggestions, and revisions; they deserve much of the credit for the evolution of the manuscript toward its present form and content. I do, however, claim all blame for its remaining failings.

Accordingly, I would like to thank Dale Amon, David Anderson, Alice Barkan, James Bennett, David Blackwell, Kenneth Boulding, Joe Boyle, Stephen Bridge, James Cataldo, Fred and Linda Chamberlain, Hugh Daniel, Douglas Denholm, Peter Diamandis, Thomas Donaldson, Allan Drexler, Hazel Drexler, Arthur Dula, Freeman Dyson, Erika Erdmann, Robert Ettinger, Mike Federowicz, Carl Feynman, David Forrest, Christopher Fry, Andy, Donna, Mark, and Scott Gassmann, Hazel and Ralph Gassmann, Agnes Gregory, Roger Gregory, David Hannah, Keith Henson, Eric Hill, Hugh Hixon, Miriam Hopkins, Joe Hopkins, Barbara Marx Hubbard, Scott A. Jones, Arthur Kantrowitz, Manfred Karnovsky, Pamela Keller, Tom and Mara Lansing, Jerome Lettvin, Elaine Lewis, David Lindbergh, Spencer Love, Robert and Susan Lovell, Steve Lubar, Arel Lucas,

John Mann, Jeff MacGillivray, Bruce Mackenzie, Marvin Minsky, Chip Morningstar, Philip Morrison, Kevin Nelson, Hugh O'Neill, Gayle Pergamit, Gordon and Mary Peterson, Norma and Amy Peterson, Naomi Reynolds, Carol Rosin, Phil Salin, Conrad Schneiker, Alice Dawn Schuster, Rosemary Simpson, Leif Smith, Ray Sperber, David Sykes, Paul Trachtman, Kevin Ulmer, Patricia Wagner, Christopher Walsh, Steve Witham, David Woodcock, and Elisa Wynn. Since this list was compiled from imperfect files and heaps of marked-up manuscripts, I apologize to those I may have omitted. Further thanks are due to the members of many audiences, at MIT and elsewhere, for asking questions that helped me refine these ideas and their presentation.

For their help and encouragement, I would also like to thank my agent, Norman Kurz, and my editors, James Raimes, Dave Barbor, and Patrick Filley. Finally, for contributions of special quality and magnitude throughout this effort, I would like to thank Mark S. Miller and, most of all, Christine Peterson. Without her help, it would not have been possible at all.

CONTENTS

PART ONE

The Foundations of Foresight

1

Engines of Construction

Protein engineering . . . represents the first major step toward a more general capability for molecular engineering which would allow us to structure matter atom by atom.

—KEVIN ULMER
Director of Exploratory Research
Genex Corporation

COAL AND DIAMONDS, sand and computer chips, cancer and healthy tissue: throughout history, variations in the arrangement of atoms have distinguished the cheap from the cherished, the diseased from the healthy. Arranged one way, atoms make up soil, air, and water; arranged another, they make up ripe strawberries. Arranged one way, they make up homes and fresh air; arranged another, they make up ash and smoke.

Our ability to arrange atoms lies at the foundation of technology. We have come far in our atom arranging, from chipping flint for arrowheads to machining aluminum for spaceships. We take pride in our technology, with our lifesaving drugs and desktop computers. Yet our spacecraft are still crude, our computers are still stupid, and the molecules in our tissues still slide into disorder, first destroying health, then life itself. For all our advances in arranging atoms, we still use primitive methods. With our present technology, we are still forced to handle atoms in unruly herds.

But the laws of nature leave plenty of room for progress, and the pressures of world competition are even now pushing us forward. For better or for worse, the greatest technological breakthrough in history is still to come.

TWO STYLES OF TECHNOLOGY

Our modern technology builds on an ancient tradition. Thirty thousand years ago, chipping flint was the high technology of the day. Our ancestors grasped stones containing trillions of trillions of atoms and removed chips containing billions of trillions of atoms to make their axheads; they made fine work with skills difficult to imitate today. They also made patterns on cave walls in France with sprayed paint, using their hands as stencils. Later they made pots by baking clay, then bronze by cooking rocks. They shaped bronze by pounding it. They made iron, then steel, and shaped it by heating, pounding, and removing chips.

We now cook up pure ceramics and stronger steels, but we still shape them by pounding, chipping, and so forth. We cook up pure silicon, saw it into slices, and make patterns on its surface using tiny stencils and sprays of light. We call the products "chips" and we consider them exquisitely small, at least in comparison to axheads.

Our microelectronic technology has managed to stuff machines as powerful as the room-sized computers of the early 1950s onto a few silicon chips in a pocket-sized computer. Engineers are now making ever smaller devices, slinging herds of atoms at a crystal surface to build up wires and components one tenth the width of a fine hair.

These microcircuits may be small by the standards of flint chippers, but each transistor still holds trillions of atoms, and so-called "microcomputers" are still visible to the naked eye. By the standards of a newer, more powerful technology they will seem gargantuan.

The ancient style of technology that led from flint chips to silicon chips handles atoms and molecules in bulk; call it *bulk technology.* The new technology will handle individual atoms and molecules with control and precision; call it *molecular technology.* It will change our world in more ways than we can imagine.

*Micro*circuits have parts measured in *micro*meters—that is, in millionths of a meter—but molecules are measured in *nano*meters (a thousand times smaller). We can use the terms "nanotechnology" and "molecular technology" interchangeably to describe the new style of technology. The engineers of the new technology will build both nanocircuits and nanomachines.

MOLECULAR TECHNOLOGY TODAY

One dictionary definition of a machine is "any system, usually of rigid bodies, formed and connected to alter, transmit, and direct applied forces in a predetermined manner to accomplish a specific objective, such as the performance of useful work." Molecular machines fit this definition quite well.

To imagine these machines, one must first picture molecules. We can picture atoms as beads and molecules as clumps of beads, like a child's beads linked by snaps. In fact, chemists do sometimes visualize molecules by building models from plastic beads (some of which link in several directions, like the hubs in a Tinkertoy set). Atoms are rounded like beads, and although molecular bonds are not snaps, our picture at least captures the essential notion that bonds can be broken and reformed.

If an atom were the size of a small marble, a fairly complex molecule would be the size of your fist. This makes a useful mental image, but atoms are really about 1/10,000 the size of bacteria, and bacteria are about 1/10,000 the size of mosquitoes. (An atomic nucleus, however, is about 1/100,000 the size of the atom itself; the difference between an atom and its nucleus is the difference between a fire and a nuclear reaction.)

The things around us act as they do because of the way their molecules behave. Air holds neither its shape nor its volume because its molecules move freely, bumping and ricocheting through open space. Water molecules stick together as they move about, so water holds a constant volume as it changes shape. Copper holds its shape because its atoms stick together in regular patterns; we can bend it and hammer it because its atoms can slip over one another while remaining bound together. Glass shatters when we hammer it because its atoms separate before they slip. Rubber consists of networks of kinked molecules, like a tangle of springs. When stretched and released, its molecules straighten and then coil again. These simple molecular patterns make up passive substances. More complex patterns make up the active nanomachines of living cells.

Biochemists already work with these machines, which are chiefly made of protein, the main engineering material of living cells. These molecular machines have relatively few atoms, and so they have lumpy surfaces, like objects made by gluing together a handful of small marbles. Also, many pairs of atoms are linked by bonds that can bend or rotate, and so protein machines are unusually flexible. But like all machines, they have parts of different shapes and sizes that do useful work. All machines use clumps of atoms as parts. Protein machines simply use very small clumps.

Biochemists dream of designing and building such devices, but there are difficulties to be overcome. Engineers use beams of light to project patterns onto silicon chips, but chemists must build much more indirectly than that. When they combine molecules in various sequences, they have only limited control over how the molecules join. When biochemists need complex molecular machines, they still have to borrow them from cells. Nevertheless, advanced molecular machines will eventually let them build nanocircuits and nanomachines as easily and directly as engineers now build microcircuits or washing machines. Then progress will become swift and dramatic.

Genetic engineers are already showing the way. Ordinarily, when chemists make molecular chains—called "polymers"—they dump molecules into a vessel where they bump and snap together haphazardly in a liquid. The resulting chains have varying lengths, and the molecules are strung together in no particular order.

But in modern gene synthesis machines, genetic engineers build more orderly polymers—specific DNA molecules—by combining molecules in a particular order. These molecules are the nucleotides of DNA (the letters of the genetic alphabet) and genetic engineers don't dump them all in together. Instead, they direct the machine to add different nucleotides in a particular sequence to spell out a particular message. They first bond one kind of nucleotide to the chain ends, then wash away the leftover material and add chemicals to prepare the chain ends to bond the next nucleotide. They grow chains as they bond on nucleotides, one at a time, in a programmed sequence. They anchor the very first nucleotide in each chain to a solid surface to keep the chain from washing away with its chemical bathwater. In this way, they have a big clumsy machine in a cabinet assemble specific molecular structures from parts a hundred million times smaller than itself.

But this blind assembly process accidentally omits nucleotides from some chains. The likelihood of mistakes grows as chains grow longer. Like workers discarding bad parts before assembling a car, genetic engineers reduce errors by discarding bad chains. Then, to join these short chains into working genes (typically thousands of nucleotides long), they turn to molecular machines found in bacteria.

These protein machines, called restriction enzymes, "read" certain DNA sequences as "cut here." They read these genetic patterns by touch, by sticking to them, and they cut the chain by rearranging a few atoms. Other enzymes splice pieces together, reading matching parts as "glue here"—likewise "reading" chains by selective stickiness and splicing chains by rearranging a few atoms. By using gene machines to write, and restriction enzymes to cut and paste, genetic engineers can write and edit whatever DNA messages they choose.

But by itself, DNA is a fairly worthless molecule. It is neither strong like Kevlar, nor colorful like a dye, nor active like an enzyme, yet it has something that industry is prepared to spend millions of dollars to use: the ability to direct molecular machines called ribosomes. In cells, molecular machines first transcribe DNA, copying its information to make RNA "tapes." Then, much as old numerically controlled machines shape metal based on instructions stored on tape, ribosomes build proteins based on instructions stored on RNA strands. And proteins are useful.

Proteins, like DNA, resemble strings of lumpy beads. But unlike DNA, protein molecules fold up to form small objects able to do things. Some are enzymes, machines that build up and tear down molecules (and copy DNA, transcribe it, and build other proteins in the cycle of life). Other proteins are hormones, binding to yet other proteins to signal cells to change their behavior. Genetic engineers can produce these objects cheaply by directing the cheap and efficient molecular machinery inside living organisms to do the work. Whereas engineers running a chemical plant must work with vats of reacting chemicals (which often misarrange atoms and make noxious byproducts), engineers working with bacteria can make them absorb chemicals, carefully rearrange the atoms, and store a product or release it into the fluid around them.

Genetic engineers have now programmed bacteria to make proteins ranging from human growth hormone to rennin, an enzyme used in making cheese. The pharmaceutical company Eli Lilly (Indianapolis) is now marketing Humulin, human insulin molecules made by bacteria.

EXISTING PROTEIN MACHINES

These protein hormones and enzymes selectively stick to other molecules. An enzyme changes its target's structure, then moves on; a hormone affects its target's behavior only so long as both remain stuck together. Enzymes and hormones can be described in mechanical terms, but their behavior is more often described in chemical terms.

But other proteins serve basic mechanical functions. Some push and pull, some act as cords or struts, and parts of some molecules make excellent bearings. The machinery of muscle, for instance, has gangs of proteins that reach, grab a "rope" (also made of protein), pull it, then reach out again for a fresh grip; whenever you move, you use these machines. Amoebas and human cells move and change shape by using fibers and rods that act as molecular muscles and bones. A reversible, variable-speed motor drives bacteria through water by turning a corkscrew-shaped propeller. If a hobbyist could build tiny cars around such motors, several billions of billions would fit in a pocket, and 150-lane freeways could be built through your finest capillaries.

Simple molecular devices combine to form systems resembling industrial machines. In the 1950s engineers developed machine tools that cut metal under the control of a punched paper tape. A century and a half earlier, Joseph-Marie Jacquard had built a loom that wove complex patterns under the control of a chain of punched cards. Yet over three billion years before Jacquard, cells had developed the machinery of the ribosome. Ribosomes are proof that nanomachines built of protein and RNA can be programmed to build complex molecules.

Then consider viruses. One kind, the T4 phage, acts like a spring-loaded syringe and looks like something out of an industrial parts catalog. It can stick to a bacterium, punch a hole, and inject viral DNA (yes, even bacteria suffer infections). Like a conqueror seizing factories to build more tanks, this DNA then directs the cell's machines to build more viral DNA and syringes. Like all organisms, these viruses exist because they are fairly stable and are good at getting copies of themselves made.

Whether in cells or not, nanomachines obey the universal laws of nature. Ordinary chemical bonds hold their atoms together, and ordinary chemical reactions (guided by other nanomachines) assemble them. Protein molecules can even join to form machines without special help, driven only by thermal agitation and chemical forces. By mixing viral proteins (and the DNA they serve) in a test tube, molecular biologists have assembled working T4 viruses. This ability is surprising: imagine putting automotive parts in a large box, shaking it, and finding an assembled car when you look inside! Yet the T4 virus is but one of many self-assembling structures. Molecular biologists have taken the machinery of the ribosome apart into over fifty separate protein and RNA molecules, and then combined them in test tubes to form working ribosomes again.

To see how this happens, imagine different T4 protein chains floating around in water. Each kind folds up to form a lump with distinctive bumps and hollows, covered by distinctive patterns of oiliness, wetness, and electric charge. Picture them wandering and tumbling, jostled by the thermal vibrations of the surrounding water molecules. From time to time two bounce together, then bounce apart. Sometimes, though, two bounce together and fit, bumps in hollows, with sticky patches matching; they then pull together and stick. In this way protein adds to protein to make sections of the virus, and sections assemble to form the whole.

Protein engineers will not need nanoarms and nanohands to assemble complex nanomachines. Still, tiny manipulators will be useful and they will be built. Just as today's engineers build machinery as complex as player pianos and robot arms from ordinary motors, bearings, and moving parts, so tomorrow's biochemists will be able to use protein molecules as motors, bearings, and moving parts to build robot arms which will themselves be able to handle individual molecules.

DESIGNING WITH PROTEIN

How far off is such an ability? Steps have been taken, but much work remains to be done. Biochemists have already mapped the structures of many proteins. With gene machines to help write DNA tapes, they can direct cells to build any protein they can design. But they still don't know how to design chains that will fold up to make proteins of the right shape and function. The forces that fold proteins are weak, and the number of plausible ways a protein might fold is astronomical, so designing a large protein from scratch isn't easy

The forces that stick proteins together to form complex machines are the same ones that fold the protein chains in the first place. The differing shapes and kinds of stickiness of amino acids—the lumpy molecular "beads" forming protein chains—make each protein chain fold up in a specific way to form an object of a particular shape. Biochemists have learned rules that suggest how an amino acid chain might fold, but the rules aren't very firm. Trying to predict how a chain will fold is like trying to work a jigsaw puzzle, but a puzzle with no pattern printed on its pieces to show when the fit is correct, and with pieces that seem to fit together about as well (or as badly) in many different ways, all but one of them wrong. False starts could consume many lifetimes, and a correct answer might not even be recognized. Biochemists using the best computer programs now available still cannot predict how a long, natural protein chain will actually fold, and some of them have despaired of designing protein molecules soon.

Yet most biochemists work as scientists, not as engineers. They work at predicting how natural proteins will fold, not at *designing* proteins that will fold predictably. These tasks may sound similar, but they differ greatly: the first is a scientific challenge, the second is an engineering challenge. Why should natural proteins fold in a way that scientists will find easy to predict? All that nature requires is that they in fact fold correctly, not that they fold in a way obvious to people.

Proteins *could* be designed from the start with the goal of making their folding more predictable. Carl Pabo, writing in the journal *Nature,* has suggested a design strategy based on this insight, and some biochemical engineers have designed and built short chains of a few dozen pieces that fold and nestle onto the surfaces of other molecules as planned. They have designed from scratch a protein with properties like those of melittin, a toxin in bee venom. They have modified existing enzymes, changing their behaviors in predictable ways. Our understanding of proteins is growing daily.

In 1959, according to biologist Garrett Hardin, some geneticists called genetic engineering impossible; today, it is an industry. Biochemistry and computer-aided design are now exploding fields, and as Frederick Blattner wrote in the journal *Science,* "computer chess programs have already reached the level below the grand master. Perhaps the solution to the protein-folding problem is nearer than we think." William Rastetter of Genentech, writing in *Applied Bio-*

chemistry and Biotechnology, asks, "How far off is *de novo* enzyme design and synthesis? Ten, fifteen years?" He answers, "Perhaps not that long."

Forrest Carter of the U.S. Naval Research Laboratory, Ari Aviram and Philip Seiden of IBM, Kevin Ulmer of Genex Corporation, and other researchers in university and industrial laboratories around the globe have already begun theoretical work and experiments aimed at developing molecular switches, memory devices, and other structures that could be incorporated into a protein-based computer. The U.S. Naval Research Laboratory has held two international workshops on molecular electronic devices, and a meeting sponsored by the U.S. National Science Foundation has recommended support for basic research aimed at developing molecular computers. Japan has reportedly begun a multimillion-dollar program aimed at developing self-assembling molecular motors and computers, and VLSI Research Inc., of San Jose, reports that "It looks like the race to bio-chips [another term for molecular electronic systems] has already started. NEC, Hitachi, Toshiba, Matsushita, Fujitsu, Sanyo-Denki and Sharp have commenced full-scale research efforts on bio-chips for bio-computers."

Biochemists have other reasons to want to learn the art of protein design. New enzymes promise to perform dirty, expensive chemical processes more cheaply and cleanly, and novel proteins will offer a whole new spectrum of tools to biotechnologists. We are already on the road to protein engineering, and as Kevin Ulmer notes in the quote from *Science* that heads this chapter, this road leads "toward a more general capability for molecular engineering which would allow us to structure matter atom by atom."

SECOND-GENERATION NANOTECHNOLOGY

Despite its versatility, protein has shortcomings as an engineering material. Protein machines quit when dried, freeze when chilled, and cook when heated. We do not build machines of flesh, hair, and gelatin; over the centuries, we have learned to use our hands of flesh and bone to build machines of wood, ceramic, steel, and plastic. We will do likewise in the future. We will use protein machines to build nanomachines of tougher stuff than protein.

As nanotechnology moves beyond reliance on proteins, it will grow more ordinary from an engineer's point of view. Molecules will be assembled like the components of an erector set, and well-bonded parts will stay put. Just as ordinary tools can build ordinary machines from parts, so molecular tools will bond molecules together to make tiny gears, motors, levers, and casings, and assemble them to make complex machines.

Parts containing only a few atoms will be lumpy, but engineers can work with lumpy parts if they have smooth bearings to support them. Conveniently enough, some bonds between atoms make fine bearings; a part can be mounted by means of a single chemical bond that will let it turn freely and smoothly. Since a bearing can be made using only two atoms (and since moving parts need have only a few atoms), nanomachines can indeed have mechanical components of molecular size.

How will these better machines be built? Over the years, engineers have used technology to improve technology. They have used metal tools to shape metal into better tools, and computers to design and program better computers. They will likewise use protein nanomachines to build better nanomachines. Enzymes show the way: they assemble large molecules by "grabbing" small molecules from the water around them, then holding them together so that a bond forms. Enzymes assemble DNA, RNA, proteins, fats, hormones, and chlorophyll in this way—indeed, virtually the whole range of molecules found in living things.

Biochemical engineers, then, will construct new enzymes to assemble new patterns of atoms. For example, they might make an enzyme-like machine which will add carbon atoms to a small spot, layer on layer. If bonded correctly, the atoms will build up to form a fine, flexible diamond fiber having over fifty times as much strength as the same weight of aluminum. Aerospace companies will line up to buy such fibers by the ton to make advanced composites. (This shows one small reason why military competition will drive molecular technology forward, as it has driven so many fields in the past.)

But the great advance will come when protein machines are able to make structures more complex than mere fibers. These programmable protein machines will resemble ribosomes programmed by RNA, or the older generation of automated machine tools programmed by punched tapes. They will open a new world of possibilities, letting engineers escape the limitations of proteins to build rugged, compact machines with straightforward designs.

Engineered proteins will split and join molecules as enzymes do.

Existing proteins bind a variety of smaller molecules, using them as chemical tools; newly engineered proteins will use all these tools and more.

Further, organic chemists have shown that chemical reactions can produce remarkable results even without nanomachines to guide the molecules. Chemists have no direct control over the tumbling motions of molecules in a liquid, and so the molecules are free to react in any way they can, depending on how they bump together. Yet chemists nonetheless coax reacting molecules to form regular structures such as cubic and dodecahedral molecules, and to form unlikely-seeming structures such as molecular rings with highly strained bonds. Molecular machines will have still greater versatility in bondmaking, because they can use similar molecular motions to make bonds, but can guide these motions in ways that chemists cannot.

Indeed, because chemists cannot yet direct molecular motions, they can seldom assemble complex molecules according to specific plans. The largest molecules they can make with specific, complex patterns are all linear chains. Chemists form these patterns (as in gene machines) by adding molecules in sequence, one at a time, to a growing chain. With only one possible bonding site per chain, they can be sure to add the next piece in the right place.

But if a rounded, lumpy molecule has (say) a hundred hydrogen atoms on its surface, how can chemists split off just one *particular* atom (the one five up and three across from the bump on the front) to add something in its place? Stirring simple chemicals together will seldom do the job, because small molecules can seldom select specific places to react with a large molecule. But protein machines will be more choosy.

A flexible, programmable protein machine will grasp a large molecule (the workpiece) while bringing a small molecule up against it in just the right place. Like an enzyme, it will then bond the molecules together. By bonding molecule after molecule to the workpiece, the machine will assemble a larger and larger structure while keeping complete control of how its atoms are arranged. This is the key ability that chemists have lacked.

Like ribosomes, such nanomachines can work under the direction of molecular tapes. Unlike ribosomes, they will handle a wide variety of small molecules (not just amino acids) and will join them to the workpiece anywhere desired, not just to the end of a chain. Protein machines will thus combine the splitting and joining abilities of enzymes with the programmability of ribosomes. But whereas ribosomes can build only the loose folds of a protein, these protein machines will build small, solid objects of metal, ceramic, or diamond—invisibly small, but rugged.

Where our fingers of flesh are likely to bruise or burn, we turn to steel tongs. Where protein machines are likely to crush or disintegrate, we will turn to nanomachines made of tougher stuff.

UNIVERSAL ASSEMBLERS

These second-generation nanomachines—built of more than just proteins—will do all that proteins can do, and more. In particular, some will serve as improved devices for assembling molecular structures. Able to tolerate acid or vacuum, freezing or baking, depending on design, enzyme-like second-generation machines will be able to use as "tools" almost any of the reactive molecules used by chemists—but they will wield them with the precision of programmed machines. They will be able to bond atoms together in virtually any stable pattern, adding a few at a time to the surface of a workpiece until a complex structure is complete. Think of such nanomachines as *assemblers*.

Because assemblers will let us place atoms in almost any reasonable arrangement (as discussed in the Notes), they will let us build almost anything that the laws of nature allow to exist. In particular, they will let us build almost anything we can design—including more assemblers. The consequences of this will be profound, because our crude tools have let us explore only a small part of the range of possibilities that natural law permits. Assemblers will open a world of new technologies.

Advances in the technologies of medicine, space, computation, and production—and warfare—all depend on our ability to arrange atoms. With assemblers, we will be able to remake our world or destroy it. So at this point it seems wise to step back and look at the prospect as clearly as we can, so we can be sure that assemblers and nanotechnology are not a mere futurological mirage.

NAILING DOWN CONCLUSIONS

In everything I have been describing, I have stuck closely to the demonstrated facts of chemistry and molecular biology. Still, people regularly raise certain questions rooted in physics and biology. These deserve more direct answers.

• *Will the uncertainty principle of quantum physics make molecular machines unworkable?*

This principle states (among other things) that particles can't be pinned down in an exact location for any length of time. It limits what molecular machines can do, just as it limits what anything else can do. Nonetheless, calculations show that the uncertainty principle places few important limits on how well atoms can be held in place, at least for the purposes outlined here. The uncertainty principle makes *electron* positions quite fuzzy, and in fact this fuzziness determines the very size and structure of atoms. An atom as a whole, however, has a comparatively definite position set by its comparatively massive nucleus. If atoms didn't stay put fairly well, molecules would not exist. One needn't study quantum mechanics to trust these conclusions, because molecular machines in the cell demonstrate that molecular machines work.

• *Will the molecular vibrations of heat make molecular machines unworkable or too unreliable for use?*

Thermal vibrations will cause greater problems than will the uncertainty principle, yet here again existing molecular machines directly demonstrate that molecular machines can work at ordinary temperatures. Despite thermal vibrations, the DNA-copying machinery in some cells makes less than one error in 100,000,000,000 operations. To achieve this accuracy, however, cells use machines (such as the enzyme DNA polymerase I) that proofread the copy and correct errors. Assemblers may well need similar error-checking and error-correcting abilities, if they are to produce reliable results.

• *Will radiation disrupt molecular machines and render them unusable?*

High-energy radiation can break chemical bonds and disrupt molecular machines. Living cells once again show that solutions exist: they operate for years by repairing and replacing radiation-damaged parts. Because individual machines are so tiny, however, they present small targets for radiation and are seldom hit. Still, if a system of nanomachines must be reliable, then it will have to tolerate a certain amount of damage, and damaged parts must regularly be repaired or replaced. This approach to reliability is well known to designers of aircraft and spacecraft.

• *Since evolution has failed to produce assemblers, does this show that they are either impossible or useless?*

The earlier questions were answered in part by pointing to the working molecular machinery of cells. This makes a simple and powerful case that natural law permits small clusters of atoms to behave as controlled machines, able to build other nanomachines. Yet despite their basic resemblance to ribosomes, assemblers will differ from anything found in cells; the things they do—while consisting of ordinary molecular motions and reactions—will have novel results. No cell, for example, makes diamond fiber.

The idea that new kinds of nanomachinery will bring new, useful abilities may seem startling: in all its billions of years of evolution, life has never abandoned its basic reliance on protein machines. Does this suggest that improvements are impossible, though? Evolution progresses through small changes, and evolution *of* DNA cannot easily *replace* DNA. Since the DNA/RNA/ribosome system is specialized to make proteins, life has had no real opportunity to evolve an alternative. Any production manager can well appreciate the reasons; even more than a factory, life cannot afford to shut down to replace its old systems.

Improved molecular machinery should no more surprise us than alloy steel being ten times stronger than bone, or copper wires transmitting signals a million times faster than nerves. Cars outspeed cheetahs, jets outfly falcons, and computers already outcalculate headscratching humans. The future will bring further examples of improvements on biological evolution, of which second-generation nanomachines will be but one.

In physical terms, it is clear enough why advanced assemblers will be able to do more than existing protein machines. They will be programmable like ribosomes, but they will be able to use a wider range of tools than all the enzymes in a cell put together. Because they will be made of materials far more strong, stiff, and stable than proteins, they will be able to exert greater forces, move with greater precision, and endure harsher conditions. Like an industrial robot arm—but unlike anything in a living cell—they will be able to rotate and move molecules in three dimensions under programmed control, making possible the precise assembly of complex objects. These advantages will enable them to assemble a far wider range of molecular structures than living cells have done.

• *Is there some special magic about life, essential to making molecular machinery work?*

One might doubt that artificial nanomachines could even equal the abilities of nanomachines in the cell, if there were reason to think that cells contained some special magic that makes them work. This idea is called "vitalism." Biologists have abandoned it because they have found chemical and physical explanations for every aspect of living cells yet studied, including their motion, growth, and reproduction. Indeed, this knowledge is the very foundation of biotechnology.

Nanomachines floating in sterile test tubes, free of cells, have been made to perform all the basic sorts of activities that they perform inside living cells. Starting with chemicals that can be made from smoggy air, biochemists have built working protein machines without help from cells. R. B. Merrifield, for example, used chemical techniques to assemble simple amino acids to make bovine pancreatic ribonuclease, an enzymatic device that disassembles RNA molecules. Life is special in structure, in behavior, and in what it feels like from the inside to be alive, yet the laws of nature that govern the machinery of life also govern the rest of the universe.

• *The case for the feasibility of assemblers and other nanomachines may sound firm, but why not just wait and see whether they can be developed?*

Sheer curiosity seems reason enough to examine the possibilities opened by nanotechnology, but there are stronger reasons. These developments will sweep the world within ten to fifty years—that is, within the expected lifetimes of ourselves or our families. What is more, the conclusions of the following chapters suggest that a wait-and-see policy would be very expensive—that it would cost many millions of lives, and perhaps end life on Earth.

Is the case for the feasibility of nanotechnology and assemblers firm enough that they should be taken seriously? It seems so, because the heart of the case rests on two well-established facts of science and engineering. These are (1) that existing molecular machines serve a range of basic functions, and (2) that parts serving these basic functions can be combined to build complex machines. Since chemical reactions can bond atoms together in diverse ways, and since molecular machines can direct chemical reactions according to programmed instructions, assemblers definitely are feasible.

NANOCOMPUTERS

Assemblers will bring one breakthrough of obvious and basic importance: engineers will use them to shrink the size and cost of computer circuits and speed their operation by enormous factors.

With today's bulk technology, engineers make patterns on silicon chips by throwing atoms and photons at them, but the patterns remain flat and molecular-scale flaws are unavoidable. With assemblers, however, engineers will build circuits in three dimensions, and build to atomic precision. The exact limits of electronic technology today remain uncertain because the quantum behavior of electrons in complex networks of tiny structures presents complex problems, some of them resulting directly from the uncertainty principle. Whatever the limits are, though, they will be reached with the help of assemblers.

The fastest computers will use electronic effects, but the smallest may not. This may seem odd, yet the essence of computation has nothing to do with electronics. A digital computer is a collection of switches able to turn one another on and off. Its switches start in one pattern (perhaps representing 2 + 2), then switch one another into a new pattern (representing 4), and so on. Such patterns can represent almost anything. Engineers build computers from tiny electrical switches connected by wires simply because mechanical switches connected by rods or strings would be big, slow, unreliable, and expensive, today.

The idea of a purely mechanical computer is scarcely new. In England during the mid-1800s, Charles Babbage invented a mechanical computer built of brass gears; his co-worker Augusta Ada, the Countess of Lovelace, invented computer programming. Babbage's endless redesigning of the machine, problems with accurate manufacturing, and opposition from budget-watching critics (some doubting the usefulness of computers!), combined to prevent its completion.

In this tradition, Danny Hillis and Brian Silverman of the MIT Artificial Intelligence Laboratory built a special-purpose mechanical computer able to play tic-tac-toe. Yards on a side, full of rotating shafts and movable frames that represent the state of the board and the strategy of the game, it now stands in the Computer Museum in Boston. It looks much like a large ball-and-stick molecular model, for it is built of Tinkertoys.

Brass gears and Tinkertoys make for big, slow computers. With components a few atoms wide, though, a simple mechanical computer would fit within 1/100 of a cubic micron, many billions of times more compact than today's so-called microelectronics. Even with a billion bytes of storage, a nanomechanical computer could fit in a box a micron wide, about the size of a bacterium. And it would be fast. Although mechanical signals move about 100,000 times slower than the electrical signals in today's machines, they will need to travel only 1/1,000,000 as far, and thus will face less delay. So a mere mechanical computer will work faster than the electronic whirlwinds of today.

Electronic nanocomputers will likely be thousands of times faster than electronic microcomputers—perhaps hundreds of thousands of times faster, if a scheme proposed by Nobel Prize-winning physicist Richard Feynman works out. Increased speed through decreased size is an old story in electronics.

DISASSEMBLERS

Molecular computers will control molecular assemblers, providing the swift flow of instructions needed to direct the placement of vast numbers of atoms. Nanocomputers with molecular memory devices will also store data generated by a process that is the opposite of assembly.

Assemblers will help engineers synthesize things; their relatives, disassemblers, will help scientists and engineers analyze things. The case for assemblers rests on the ability of enzymes and chemical reactions to form bonds, and of machines to control the process. The case for disassemblers rests on the ability of enzymes and chemical reactions to break bonds, and of machines to control the process. Enzymes, acids, oxidizers, alkali metals, ions, and reactive groups of atoms called free radicals—all can break bonds and remove groups of atoms. Because nothing is absolutely immune to corrosion, it seems that molecular tools will be able to take anything apart, a few atoms at a time. What is more, a nanomachine could (at need or convenience) apply mechanical force as well, in effect prying groups of atoms free.

A nanomachine able to do this, while recording what it removes layer by layer, is a *disassembler*. Assemblers, disassemblers, and nanocomputers will work together. For example, a nanocomputer system will be able to direct the disassembly of an object, record its structure, and then direct the assembly of perfect copies. And this gives some hint of the power of nanotechnology.

THE WORLD MADE NEW

Assemblers will take years to emerge, but their emergence seems almost inevitable: Though the path to assemblers has many steps, each step will bring the next in reach, and each will bring immediate rewards. The first steps have already been taken, under the names of "genetic engineering" and "biotechnology." Other paths to assemblers seem possible. Barring worldwide destruction or worldwide controls, the technology race will continue whether we wish it or not. And as advances in computer-aided design speed the development of molecular tools, the advance toward assemblers will quicken.

To have any hope of understanding our future, we must understand the consequences of assemblers, disassemblers, and nanocomputers. They promise to bring changes as profound as the industrial revolution, antibiotics, and nuclear weapons all rolled up in one massive breakthrough. To understand a future of such profound change, it makes sense to seek principles of change that have survived the greatest upheavals of the past. They will prove a useful guide.

2

The Principles of Change

Think of the design process as involving first the generation of
alternatives and then the testing of these alternatives against a whole
array of requirements and constraints.

—HERBERT A. SIMON

MOLECULAR ASSEMBLERS will bring a revolution without parallel since the development of ribosomes, the primitive assemblers in
the cell. The resulting nanotechnology can help life spread beyond
Earth—a step without parallel since life spread beyond the seas. It
can help mind emerge in machines—a step without parallel since
mind emerged in primates. And it can let our minds renew and
remake our bodies—a step without any parallel at all.

These revolutions will bring dangers and opportunities too vast
for the human imagination to grasp. Yet the principles of change that
have applied to molecules, cells, beasts, minds, and machines should
endure even in an age of biotechnology, nanomachines, and artificial
minds. The same principles that have applied at sea, on land, and in
the air should endure as we spread Earth's life toward the stars.
Understanding the enduring principles of change will help us understand the potential for good and ill in the new technologies.

ORDER FROM CHAOS

Order can emerge from chaos without anyone's giving orders: orderly crystals condensed from formless interstellar gas long before Sun, Earth, or life appeared. Chaos also gives rise to a crystalline order under more familiar circumstances. Imagine a molecule—perhaps regular in form, or perhaps lopsided and knobby like a ginger root. Now imagine a vast number of such molecules moving randomly in a liquid, tumbling and jostling like drunkards in weightlessness in the dark. Imagine the liquid evaporating and cooling, forcing the molecules closer together and slowing them down. Will these randomly moving, oddly shaped molecules simply gather in disordered heaps? Generally not. They will usually settle into a crystalline pattern, each neatly nestled against its neighbors, forming rows and columns as perfect as a checkerboard, though often more complex.

This process involves neither magic nor some special property of molecules and quantum mechanical forces. It does not even require the special matching shapes that enable protein molecules to self-assemble into machines. Marbles of uniform size, if placed in a tray and shaken, also settle into a regular pattern.

Crystals grow by trial and the removal of error, by variation and selection. No tiny hands assemble them. A crystal can begin with a chance clumping of molecules: the molecules wander, bump, and clump at random, but clumps stick best when packed in the right crystalline pattern. Other molecules then strike this first, tiny crystal. Some bump in the wrong position or orientation; they stick poorly and shake loose again. Others happen to bump properly; they stick better and often stay. Layer builds on layer, extending the crystalline pattern. Though the molecules bump at random, they do not stick at random. Order grows from chaos through variation and selection.

EVOLVING MOLECULES

In crystal growth, each layer forms a template for the next. Uniform layers accumulate to form a solid block.

In cells, strands of DNA or RNA can serve as templates too, aided by enzymes that act as molecular copying machines. But the subunits of nucleic acid strands can be arranged in many different sequences, and a template strand can separate from its copy. Both strand and copy can then be copied again. Biochemist Sol Spiegelman has used a copying machine (a protein from a virus) in test tube experiments. In a simple, lifeless environment, it duplicates RNA molecules.

Picture a strand of RNA floating in a test tube together with copying machines and RNA subunits. The strand tumbles and writhes until it bumps into a copying machine in the right position to stick. Subunits bump around until one of the right kind meets the copying machine in the right position to match the template strand. As matching subunits chance to fall into position, the machine seizes them and bonds them to the growing copy; though subunits bump randomly, the machine bonds selectively. Finally the machine, the template, and the copy separate.

In the terminology of Oxford zoologist Richard Dawkins, things that give rise to copies of themselves are called *replicators*. In this environment, RNA molecules qualify: a single molecule soon becomes two, then four, eight, sixteen, thirty-two, and so forth, multiplying exponentially. Later, the replication rate levels off: the fixed stock of protein machines can churn out RNA copies only so fast, no matter how many template molecules vie for their services. Later still, the raw materials for making RNA molecules become scarce and replication starves to a halt. The exploding population of molecules reaches a limit to growth and stops reproducing.

The copying machines, however, often miscopy an RNA strand, inserting, deleting, or mismatching a subunit. The resulting mutated strand then differs in length or subunit sequence. Such changes are fairly random, and changes accumulate as miscopied molecules are again miscopied. As the molecules proliferate, they begin to grow different from their ancestors and from each other. This might seem a recipe for chaos.

Biochemists have found that differing RNA molecules replicate at differing rates, depending on their lengths and subunit patterns. Descendants of the swifter replicators naturally grow more common. Indeed, if one kind replicates just 10 percent more rapidly than its siblings, then after one hundred generations, each of the faster kind gives rise to 1,000 times as many descendants. Small differences in exponential growth pile up exponentially.

When a test tube runs out of subunits, an experimenter can sample its RNA and "infect" a fresh tube. The process begins again and the molecules that dominated the first round of competition begin with a head start. More small changes appear, building over time into large changes. Some molecules replicate faster, and their kind dominates the mix. When resources run out, the experimenter can sample the RNA and start again (and again, and again), holding conditions stable.

This experiment reveals a natural process: no matter what RNA sequences the experimenter starts with, the seeming chaos of random errors and biased copying brings forth one kind of RNA molecule (give or take some copying errors). Its typical version has a known, well-defined sequence of 220 subunits. It is the best RNA replicator in this environment, so it crowds out the others and stays.

Prolonged copying, miscopying, and competition always bring about the same result, no matter what the length or pattern of the RNA molecule that starts the process. Though no one could have predicted this winning pattern, anyone can see that change and competition will tend to bring forth a single winner. Little else could happen in so simple a system. If these replicators affected one another strongly (perhaps by selectively attacking or helping one another), then the result could resemble a more complex ecology. As it is, they just compete for a resource.

A variation on this example shows us something else: RNA molecules adapt differently to different environments. A molecular machine called a ribonuclease grabs RNA molecules having certain sequences of exposed subunits and cuts them in two. But RNA molecules, like proteins, fold in patterns that depend on their sequences, and by folding the right way they can protect their vulnerable spots. Experimenters find that RNA molecules evolve to sacrifice swift replication for better protection when ribonuclease is around. Again, a best competitor emerges.

Notice that biological terms have crept into this description: since the molecules replicate, the word "generation" seems right; molecules "descended" from a common "ancestor" are "relatives," and the words "growth," "reproduction," "mutation," and "competition" also seem right. Why is this? Because these molecules copy themselves with small variations, as do the genes of living organisms. When varying replicators have varying successes, the more successful tend to accumulate. This process, wherever it occurs, is "evolution."

In this test tube example we can see evolution stripped to its bare essentials, free of the emotional controversy surrounding the evolution of life. The RNA replicators and protein copying machines are well-defined collections of atoms obeying well-understood principles and evolving in repeatable laboratory conditions. Biochemists can make RNA and protein from off-the-shelf chemicals, without help from life.

Biochemists borrow these copying machines from a kind of virus that infects bacteria and uses RNA as its genetic material. These viruses survive by entering a bacterium, getting themselves copied using its resources, and then escaping to infect new bacteria. Miscopying of viral RNA produces mutant viruses, and viruses that replicate more successfully grow more common; this is evolution by natural selection, apparently called "natural" because it involves nonhuman parts of nature. But unlike the test tube RNA, viral RNA must do more than just replicate itself as a bare molecule. Successful viral RNA must also direct bacterial ribosomes to build protein devices that let it first escape from the old bacterium, then survive outside, and finally enter a new one. This additional information makes viral RNA molecules about 4,500 subunits long.

To replicate successfully, the DNA of large organisms must do even more, directing the construction of tens of thousands of different protein machines and the development of complex tissues and organs. This requires thousands of genes coded in millions to billions of DNA subunits. Nevertheless, the essential process of evolution by variation and selection remains the same in the test tube, in viruses, and far beyond.

EXPLAINING ORDER

There are at least three ways to explain the structure of an evolved population of molecular replicators, whether test tube RNA, viral genes, or human genes. The first kind of explanation is a blow-by-blow account of their histories: how specific mutations occurred and how they spread. This is impossible without recording all the molecular events, and such a record would in any event be immensely tedious.

The second kind of explanation resorts to a somewhat misleading word: purpose. In detail, the molecules simply change haphazardly and replicate selectively. Yet stepping back from the process, one *could* describe the outcome by imagining that the surviving molecules have changed to "achieve the goal" of replication. Why do RNA molecules that evolved under the threat of ribonuclease fold as they do? Because of a long and detailed history, of course, but the idea that "they want to avoid attack and survive to replicate" would predict the same result. The language of purpose makes useful shorthand (try discussing human action without it!), but the appearance of purpose need not result from the action of a mind. The RNA example shows this quite neatly.

The third (and often best) kind of explanation—in terms of evolution—says that order emerges through the variation and selection of replicators. A molecule folds in a particular way because it resembles ancestors that multiplied more successfully (by avoiding attack, etc.), and left descendants including itself. As Richard Dawkins points out, the language of purpose (if used carefully) can be translated into the language of evolution.

Evolution attributes patterns of success to the *elimination* of *unsuccessful* changes. It thus explains a positive as the result of a double negative—an explanation of a sort that seems slightly difficult to grasp. Worse, it explains something *visible* (successful, purposeful entities) in terms of something *invisible* (unsuccessful entities that have vanished). Because only successful beasts have littered the landscape with the bones of their descendants, the malformed failures of the past haven't even left many fossils.

The human mind tends to focus on the visible, seeking positive causes for positive results, an ordering force behind orderly results. Yet through reflection we can see that this great principle has changed our past and will shape our future: *Evolution proceeds by the variation and selection of replicators.*

EVOLVING ORGANISMS

The history of life is the history of an arms race based on molecular machinery. Today, as this race approaches a new and swifter phase, we need to be sure we understand just how deeply rooted evolution is. In a time when the idea of biological evolution is often slighted in the schools and sometimes attacked, we should remember that the supporting evidence is as solid as rock and as common as cells.

In pages of stone, the Earth itself has recorded the history of life. On lake bottoms and seabed, shells, bones, and silt have piled, layer on layer. Sometimes a shifting current or a geological upheaval has washed layers away; otherwise they have simply deepened. Early layers, buried deep, have been crushed, baked, soaked in mineral waters, and turned to stone.

For centuries, geologists have studied rocks to read Earth's past. Long ago, they found seashells high in the crushed and crumpled rock of mountain ranges. By 1785—seventy-four years before Darwin's detested book—James Hutton had concluded that seabed mud had been pressed to stone and raised skyward by forces not yet understood. What else could geologists think, unless nature itself had lied?

They saw that fossil bones and shells differed from layer to layer. They saw that shells in layers *here* matched shells in layers *there*, though the layers might lie deep beneath the land between. They named layers *(A,B,C,D . . .* , or Osagian, Meramecian, Lower Chesterian, Upper Chesterian . . .), and used characteristic fossils to trace rock layers. The churning of Earth's crust has nowhere left a complete sequence of layers exposed, yet geologists finding *A,B,C,D,E* in one place, *C,D,E,F,G,H,I,J* in another and *J,K,L* somewhere else could see that *A* preceded *L*. Petroleum geologists (even those who care nothing for evolution or its implications) still use such fossils to date rock layers and to trace layers from one drill site to another.

Scientists came to the obvious conclusion. Just as sea species today live in broad areas, so did species in years gone by. Just as layer piles on top of layer today, so did they then. Similar shells in similar layers mark sediments laid down in the same age. Shells change from layer to layer because species changed from age to age. This is what geologists found written in shells and bones on pages of stone.

The uppermost layers of rock contain bones of recent animals, deeper layers contain bones of animals now extinct. Still earlier layers show no trace of any modern species. Below mammal bones lie dinosaur bones; in older layers lie amphibian bones, then shells and fish bones, and then no bones or shells at all. The oldest fossil-bearing rocks bear the microscopic traces of single cells.

Radioactive dating shows these oldest traces to be several billion years old. Cells more complex than bacteria date to little more than one billion years ago. The history of worms, fish, amphibians, reptiles, and mammals spans hundreds of millions of years. Human-like bones date back several million years. The remains of civilizations date back several thousand.

In three billion years, life evolved from single cells able to soak up chemicals to collections of cells embodying minds able to soak up ideas. Within the last century, technology has evolved from the steam locomotive and electric light to the spaceship and the electronic computer—and computers are already being taught to read and write. With mind and technology, the rate of evolution has jumped a millionfold or more.

Another Route Back

The book of stone records the forms of long-dead organisms, yet living cells also carry records, genetic texts only now being read. As with the ideas of geology, the essential ideas of evolution were known before Darwin had set pen to paper.

In lamp-lit temples and monasteries, generations of scribes copied and recopied manuscripts. Sometimes they miscopied words and sentences—whether by accident, by perversity, or by order of the local ruler—and as the manuscripts replicated, aided by these human copying machines, errors accumulated. The worst errors might be caught and removed, and famous passages might survive unchanged, but differences grew.

Ancient books seldom exist in their original versions. The oldest copies are often centuries younger than the lost originals. Nonetheless, from differing copies with differing errors, scholars can reconstruct versions closer to the original.

They compare texts. They can trace lines of descent from common ancestors because unique patterns of errors betray copying from a common source. (Schoolteachers know this: identical right answers aren't a tipoff—unless on an essay test—but woe to students sitting side by side who turn in tests with identical mistakes!) Where all surviving copies agree, scholars can assume that the original copy (or at least the last shared ancestor of the survivors) held the same words. Where survivors differ, scholars study copies that descended separately from a distant ancestor, because areas of agreement then indicate a common origin in the ancestral version.

Genes resemble manuscripts written in a four-letter alphabet. Much as a message can take many forms in ordinary language (restating an idea using entirely different words is no great strain), so different genetic wording can direct the construction of identical protein molecules. Moreover, protein molecules with different design details can serve identical functions. A collection of genes in a cell is like a whole book, and genes—like old manuscripts—have been copied and recopied by inaccurate scribes.

Like scholars studying ancient texts, biologists generally work with modern copies of their material (with, alas, no biological Dead Sea Scrolls from the early days of life). They compare organisms with similar appearances (lions and tigers, horses and zebras, rats and mice) and find that they give similar answers to the essay questions in their genes and proteins. The more two organisms differ (lions and lizards, humans and sunflowers), the more these answers differ, even among molecular machines serving identical functions. More telling still, similar animals make the same mistakes—all primates, for example, lack enzymes for making vitamin C, an omission shared by only two other known mammals, the guinea pig and the fruit bat. This suggests that we primates have copied our genetic answers from a shared source, long ago.

The same principle that shows the lines of descent of ancient texts (and that helps correct their copying errors) thus also reveals the lines of descent of modern life. Indeed, it indicates that all known life shares a common ancestor.

The Rise of the Replicators

The first replicators on Earth evolved abilities beyond those possible to RNA molecules replicating in test tubes. By the time they reached the bacterial stage, they had developed the "modern" system of using DNA, RNA, and ribosomes to construct protein. Mutations then changed not only the replicating DNA itself, but protein machines and the living structures they build and shape.

Teams of genes shaped ever more elaborate cells, then guided the cellular cooperation that formed complex organisms. Variation and selection favored teams of genes that shaped beasts with protective skins and hungry mouths, animated by nerve and muscle, guided by eye and brain. As Richard Dawkins puts it, genes built ever more elaborate survival machines to aid their own replication.

When dog genes replicate, they often shuffle with those of other dogs that have been selected by people, who then select which puppies to keep and breed. Over the millennia, people have molded wolf-like beasts into greyhounds, toy poodles, dachshunds, and Saint Bernards. By selecting which genes survive, people have reshaped dogs in both body and temperament. Human desires have defined success for dog genes; other pressures have defined success for wolf genes.

Mutation and selection of genes has, through long ages, filled the world with grass and trees, with insects, fish, and people. More recently, other things have appeared and multiplied—tools, houses, aircraft, and computers. And like the lifeless RNA molecules, this hardware has *evolved*.

EVOLVING TECHNOLOGY

As the stone of Earth records the emergence of ever more complex and capable forms of life, so the relics and writings of humanity record the emergence of ever more complex and capable forms of hardware. Our oldest surviving hardware is itself stone, buried with the fossils of our ancestors; our newest hardware orbits overhead.

Consider for a moment the hybrid ancestry of the space shuttle. On its aircraft side, it descends from the aluminum jets of the sixties, which themselves sprang from a line stretching back through the aluminum prop planes of World War II, to the wood-and-cloth biplanes of World War I, to the motorized gliders of the Wright brothers, to toy gliders and kites. On its rocket side, the shuttle traces back to Moon rockets, to military missiles, to last century's artillery rockets ("and the rocket's red glare . . ."), and finally to fireworks and toys. This aircraft/rocket hybrid flies, and by varying components and designs, aerospace engineers will evolve still better ones.

Engineers speak of "generations" of technology; Japan's "fifth-generation" computer project shows how swiftly some technologies grow and spawn. Engineers speak of "hybrids," of "competing technologies," and of their "proliferation." IBM Director of Research Ralph E. Gomory emphasizes the evolutionary nature of technology, writing that "technology development is much more evolutionary and much less revolutionary or breakthrough-oriented than most people imagine." (Indeed, even breakthroughs as important as molecular assemblers will develop through many small steps.) In the quote that heads this chapter, Professor Herbert A. Simon of Carnegie-Mellon University urges us to "think of the design process as involving first the generation of alternatives and then the testing of these alternatives against a whole array of requirements and constraints." Generation and testing of alternatives is synonymous with variation and selection.

Sometimes various alternatives already exist. In "One Highly Evolved Toolbox," in *The Next Whole Earth Catalog,* J. Baldwin writes: "Our portable shop has been evolving for about twenty years now. There's nothing really very special about it except that a continuing process of removing obsolete or inadequate tools and replacing them with more suitable ones has resulted in a collection that has become a thing-making system rather than a pile of hardware."

Baldwin uses the term "evolving" accurately. Invention and manufacture have for millennia generated variations in tool designs, and Baldwin has winnowed the current crop by competitive selection, keeping those that work best with his other tools to serve his needs. Through years of variation and selection, his system evolved—a process he highly recommends. Indeed, he urges that one never try to plan out the purchase of a complete set of tools. Instead, he urges buying the tools one often borrows, tools selected not by theory but by experience.

Technological variations are often deliberate, in the sense that engineers are paid to invent and test. Still, some novelties are sheer accident, like the discovery of a crude form of Teflon in a cylinder supposedly full of tetrafluoroethylene gas: with its valve open, it remained heavy; when it was sawed open, it revealed a strange, waxy solid. Other novelties have come from systematic blundering. Edison tried carbonizing everything from paper to bamboo to spiderwebs when he was seeking a good light-bulb filament. Charles Goodyear messed around in a kitchen for years, trying to convert gummy natural rubber into a durable substance, until at last he chanced to drop sulfurized rubber on a hot stove, performing the first crude vulcanization.

In engineering, enlightened trial and error, not the planning of flawless intellects, has brought most advances; this is why engineers build prototypes. Peters and Waterman in their book *In Search of Excellence* show that the same holds true of advances in corporate products and policies. This is why excellent companies create "an environment and a set of attitudes that encourage experimentation," and why they evolve "in a very Darwinian way."

Factories bring order through variation and selection. Crude quality-control systems test and discard faulty parts before assembling products, and sophisticated quality-control systems use statistical methods to track defects to their sources, helping engineers change the manufacturing process to minimize defects. Japanese engineers, building on W. Edwards Deming's work in statistical quality control, have made such variation and selection of industrial processes a pillar of their country's economic success. Assembler-based systems will likewise need to measure results to eliminate flaws.

Quality control is a sort of evolution, aiming not at change but at eliminating harmful variations. But just as Darwinian evolution can preserve and spread favorable mutations, so good quality control systems can help managers and workers to preserve and spread more effective processes, whether they appear by accident or by design.

All this tinkering by engineers and manufacturers prepares products for their ultimate test. Out in the market, endless varieties of wrench, car, sock, and computer compete for the favor of buyers. When informed buyers are free to choose, products that do too little or cost too much eventually fail to be re-produced. As in nature, competitive testing makes yesterday's best competitor into tomorrow's fossil. "Ecology" and "economy" share more than linguistic roots.

Both in the marketplace and on real and imaginary battlefields, global competition drives organizations to invent, buy, beg, and steal ever more capable technologies. Some organizations compete chiefly to serve people with superior goods, others compete chiefly to intimidate them with superior weapons. The pressures of evolution drive both.

The global technology race has been accelerating for billions of years. The earthworm's blindness could not block the development of sharp-eyed birds. The bird's small brain and clumsy wings could not block the development of human hands, minds, and shotguns. Likewise, local prohibitions cannot block advances in military and commercial technology. It seems that we must guide the technology race or die, yet the force of technological evolution makes a mockery of anti-technology movements: democratic movements for local restraint can only restrain the world's democracies, not the world as a whole. The history of life and the potential of new technology suggest some solutions, but this is a matter for Part Three.

THE EVOLUTION OF DESIGN

It might seem that design offers an alternative to evolution, but design involves evolution in two distinct ways. First, design practice itself evolves. Not only do engineers accumulate designs that work, they accumulate design methods that work. These range from handbook standards for choosing pipes to management systems for organizing research and development. And as Alfred North Whitehead stated, "The greatest invention of the nineteenth century was the invention of the method of invention."

Second, design itself proceeds by variation and selection. Engineers often use mathematical laws evolved to describe (for example) heat flow and elasticity to test *simulated* designs before building them. They thus evolve plans through a cycle of design, calculation, criticism, and redesign, avoiding the expense of cutting metal. The creation of designs thus proceeds through a nonmaterial form of evolution.

Hooke's law, for example, describes how metal bends and stretches: deformation is proportional to the applied stress: twice the pull, twice the stretch. Though only roughly correct, it remains fairly accurate until the metal's springiness finally yields to stress. Engineers can use a form of Hooke's law to design a bar of metal that can support a load without bending too far—and then make it just a bit thicker to allow for inaccuracies in the law and in their design calculations. They can also use a form of Hooke's law to describe the bending and twisting of aircraft wings, tennis rackets, and automobile frames. But simple mathematical equations don't wrap smoothly around such convoluted structures. Engineers have to fit the equations to simpler shapes (to pieces of the design), and then assemble these partial solutions to describe the flexing of the whole. It is a method (called "finite element analysis") that typically requires immense calculations, and without computers it would be impractical. With them, it has grown common.

Such simulations extend an ancient trend. We have always imagined consequences, in hope and fear, when we have needed to select a course of action. Simpler mental models (whether inborn or learned) undoubtedly guide animals as well. When based on accurate mental models, thought experiments can replace more costly (or even deadly) physical experiments—a development evolution has favored. Engineering simulations simply extend this ability to imagine consequences, to make our mistakes in thought rather than deed.

In "One Highly Evolved Toolbox," J. Baldwin discusses how tools and thought mesh in job-shop work: "You begin to build your tool capability into the way you think about making things. As anyone who makes a lot of stuff will tell you, the tools soon become sort of an automatic part of the design process . . . But tools can't become part of your design process if you don't know what is available and what the various tools do."

Having a feel for tool capabilities is essential when planning a jobshop project for delivery next Wednesday; it is equally essential when shaping a strategy for handling the breakthroughs of the coming decades. The better our feel for the future's tools, the sounder will be our plans for surviving and prospering.

A craftsman in a job shop can keep tools in plain sight; working with them every day makes them familiar to his eyes, hands, and mind. He gets to know their abilities naturally, and can put this knowledge to immediate creative use. But people—like us—who have to understand the future face a greater challenge, because the future's tools exist now only as ideas and as possibilities implicit in natural law. These tools neither hang on the wall nor impress themselves on the mind through sight and sound and touch—nor will they, until they exist as hardware. In the coming years of preparation only study, imagination, and thought can make their abilities real to the mind.

WHAT ARE THE NEW REPLICATORS?

History shows us that hardware evolves. Test tube RNA, viruses, and dogs all show how evolution proceeds by the modification and testing of replicators. But hardware (today) cannot reproduce itself —so where are the replicators behind the evolution of technology? What are the machine genes?

Of course, we need not actually identify replicators in order to recognize evolution. Darwin described evolution before Mendel discovered genes, and geneticists learned much about heredity before Watson and Crick discovered the structure of DNA. Darwin needed no knowledge of molecular genetics to see that organisms varied and that some left more descendants.

A replicator is a pattern that can get copies of itself made. It may need help; without protein machines to copy it, DNA could not replicate. But by this standard, some *machines* are replicators! Companies often make machines that fall into the hands of a competitor; the competitor then learns their secrets and builds copies. Just as genes "use" protein machines to replicate, so such machines "use" human minds and hands to replicate. With nanocomputers directing assemblers and disassemblers, the replication of hardware could even be automated.

The human mind, though, is a far subtler engine of imitation than any mere protein machine or assembler. Voice, writing, and drawing can transmit designs from mind to mind before they take form as hardware. The ideas behind methods of design are subtler yet: more abstract than hardware, they replicate and function exclusively in the world of minds and symbol systems.

Where genes have evolved over generations and eons, mental replicators now evolve over days and decades. Like genes, ideas split, combine, and take multiple forms (genes can be transcribed from DNA to RNA and back again; ideas can be translated from language to language). Science cannot yet describe the neural patterns that embody ideas in brains, but anyone can see that ideas mutate, replicate, and compete. Ideas evolve.

Richard Dawkins calls bits of replicating mental patterns "memes" (*meme* rhymes with *cream*). He says "examples of memes are tunes, ideas, catch-phrases, clothes fashions, ways of making pots or of building arches. Just as genes propagate themselves in the gene pool by leaping from body to body [generation to generation] via sperms or eggs, so memes propagate themselves in the meme pool by leaping from brain to brain via a process which, in the broad sense, can be called imitation."

THE CREATURES OF THE MIND

Memes replicate because people both learn and teach. They vary because people create the new and misunderstand the old. They are selected (in part) because people don't believe or repeat everything they hear. As test tube RNA molecules compete for scarce copying machines and subunits, so memes must compete for a scarce resource —human attention and effort. Since memes shape behavior, their success or failure is a deadly serious matter.

Since ancient times, mental models and patterns of behavior have passed from parent to child. Meme patterns that aid survival and reproduction have tended to spread. (Eat this root only after cooking; don't eat those berries, their evil spirits will twist your guts.) Year by year, people varied their actions with varying results. Year by year, some died while others found new tricks of survival and passed them on. Genes built brains skilled at imitation because the patterns imitated were, on the whole, of value—their bearers, after all, had survived to spread them.

Memes themselves, though, face their own matters of "life" and "death": as replicators, they evolve solely to survive and spread. Like viruses, they can replicate without aiding their host's survival or well-being. Indeed, the meme for martyrdom-in-a-cause can spread itself through the very act of killing its host.

Genes, like memes, survive by many strategies. Some duck genes have spread themselves by encouraging ducks to pair off to care for their gene-bearing eggs and young. Some duck genes have spread themselves (when in male ducks) by encouraging rape, and some (when in female ducks) by encouraging the planting of eggs in other ducks' nests. Still other genes found in ducks are virus genes, able to spread without making more ducks. Protecting eggs helps the duck species (and the individual duck genes) survive; rape helps one set of duck genes at the expense of others; infection helps viral genes at the expense of duck genes in general. As Richard Dawkins points out, genes "care" only about their own replication: they appear selfish.

But selfish motives can encourage cooperation. People seeking money and recognition for themselves cooperate to build corporations that serve other people's wants. Selfish genes cooperate to build organisms that themselves often cooperate. Even so, to imagine that genes *automatically* serve some greater good (—of their chromosome? —their cell? —their body? —their species?) is to mistake a common effect for an underlying cause. To ignore the selfishness of replicators is to be lulled by a dangerous illusion.

Some genes in cells are out-and-out parasites. Like herpes genes inserted in human chromosomes, they exploit cells and harm their hosts. Yet if genes can be parasites, why not memes as well?

In *The Extended Phenotype,* Richard Dawkins describes a worm that parasitizes bees and completes its life cycle in water. It gets from bee to water by making the host bee dive to its death. Similarly, ant brainworms must enter a sheep to complete their life cycle. To accomplish this, they burrow into the host ant's brain, somehow causing changes that make the ant "want" to climb to the top of a grass stem and wait, eventually to be eaten by a sheep.

As worms enter other organisms and use them to survive and replicate, so do memes. Indeed, the *absence* of memes exploiting people for their own selfish ends would be amazing, a sign of some powerful—indeed, nearly perfect—mental immune system. But parasitic memes clearly do exist. Just as viruses evolve to stimulate cells to make viruses, so rumors evolve to sound plausible and juicy, stimulating repetition. Ask not whether a rumor is true, ask instead how it spreads. Experience shows that ideas evolved to be successful replicators need have little to do with the truth.

At best, chain letters, spurious rumors, fashionable lunacies, and other mental parasites harm people by wasting their time. At worst, they implant deadly misconceptions. These meme systems exploit human ignorance and vulnerability. Spreading them is like having a cold and sneezing on a friend. Though some memes act much like viruses, infectiousness isn't necessarily bad (think of an infectious grin, or infectious good nature). If a package of ideas has merit, then its infectiousness simply increases its merit—and indeed, the best ethical teachings also teach us to teach ethics. Good publications may entertain, enrich understanding, aid judgment—and advertise gift subscriptions. Spreading useful meme systems is like offering useful seeds to a friend with a garden.

SELECTING IDEAS

Parasites have forced organisms to evolve immune systems, such as the enzymes that bacteria use to cut up invading viruses, or the roving white blood cells our bodies use to destroy bacteria. Parasitic memes have forced minds down a similar path, evolving meme systems that serve as mental immune systems.

The oldest and simplest mental immune system simply commands "believe the old, reject the new." Something like this system generally kept tribes from abandoning old, tested ways in favor of wild new notions—such as the notion that obeying alleged ghostly orders to destroy all the tribe's cattle and grain would somehow bring forth a miraculous abundance of food and armies of ancestors to drive out foreigners. (This meme package infected the Xhosa people of southern Africa in 1856; by the next year 68,000 had died, chiefly of starvation.)

Your body's immune system follows a similar rule: it generally accepts all the cell types present in early life and rejects new types, such as potential cancer cells and invading bacteria, as foreign and dangerous. This simple reject-the-new system once worked well, yet in this era of organ transplantation it can kill. Similarly, in an era when science and technology regularly present facts that are both new and trustworthy, a rigid mental immune system becomes a dangerous handicap.

For all its shortcomings, though, the reject-the-new principle is simple and offers real advantages. Tradition holds much that is tried and true (or if not true, then at least workable). Change is risky: just as most mutations are bad, so most new ideas are wrong. Even reason can be dangerous: if a tradition links sound practices to a fear of ghosts, then overconfident rational thought may throw out the good with the bogus. Unfortunately, traditions evolved to *be* good may have less appeal than ideas evolved to *sound* good—when first questioned, the soundest tradition may be displaced by worse ideas that better appeal to the rational mind.

Yet memes that seal the mind against new ideas protect themselves in a suspiciously self-serving way. While protecting valuable traditions from clumsy editing, they may also shield parasitic claptrap from the test of truth. In times of swift change they can make minds dangerously rigid.

Much of the history of philosophy and science may be seen as a search for better mental immune systems, for better ways to reject the false, the worthless, and the damaging. The best systems respect tradition, yet encourage experiment. They suggest standards for judging memes, helping the mind distinguish between parasites and tools.

The principles of evolution provide a way to view change, whether in molecules, organisms, technologies, minds, or cultures. The same basic questions keep arising: What are the replicators? How do they vary? What determines their success? How do they defend against invaders? These questions will arise again when we consider the consequences of the assembler revolution, and yet again when we consider how society might deal with those consequences.

The deep-rooted principles of evolutionary change will shape the development of nanotechnology, even as the distinction between hardware and life begins to blur. These principles show much about what we can and cannot hope to achieve, and they can help us focus our efforts to shape the future. They also tell us much about what we can and cannot foresee, because they guide the evolution not only of hardware, but of knowledge itself.

3

Predicting and Projecting

The critical attitude may be described as the conscious attempt to make our theories, our conjectures, suffer in our stead in the struggle for the survival of the fittest. It gives us a chance to survive the elimination of an inadequate hypothesis—when a more dogmatic attitude would eliminate it by eliminating us.

—Sir KARL POPPER

AS WE LOOK FORWARD to see where the technology race leads, we should ask three questions. What is *possible,* what is *achievable,* and what is *desirable?*

First, where hardware is concerned, natural law sets limits to the possible. Because assemblers will open a path to those limits, understanding assemblers is a key to understanding what is possible.

Second, the principles of change and the facts of our present situation set limits to the achievable. Because evolving replicators will play a basic role, the principles of evolution are a key to understanding what will be achievable.

As for what is desirable or undesirable, our differing dreams spur a quest for a future with room for diversity, while our shared fears spur a quest for a future of safety.

These three questions—of the possible, the achievable, and the desirable—frame an approach to foresight. First, scientific and engineering knowledge form a map of the limits of the possible. Though still blurred and incomplete, this map outlines the permanent limits within which the future must move. Second, evolutionary principles determine what paths lie open, and set limits to achievement—including lower limits, because advances that promise to improve life or to further military power will be virtually unstoppable. This allows a limited prediction: If the eons-old evolutionary race does not somehow screech to a halt, then competitive pressures will mold our technological future to the contours of the limits of the possible. Finally, within the broad confines of the possible and the achievable, we can try to reach a future we find desirable.

PITFALLS OF PROPHECY

But how can anyone predict the future? Political and economic trends are notoriously fickle, and sheer chance rolls dice across continents. Even the comparatively steady advance of technology often eludes prediction.

Prognosticators often guess at the times and costs required to harness new technologies. When they reach beyond outlining possibilities and attempt accurate predictions, they generally fail. For example, though the space shuttle was clearly possible, predictions of its cost and initial launch date were wrong by several years and billions of dollars. Engineers cannot accurately predict when a technology will be developed, because development always involves uncertainties.

But we have to try to predict and guide development. Will we develop monster technologies before cage technologies, or after? Some monsters, once loosed, cannot be caged. To survive, we must keep control by speeding some developments and slowing others.

Though one technology can sometimes block the dangers of another (defense vs. offense, pollution controls vs. pollution), competing technologies often go in the same direction. On December 29, 1959, Richard Feynman (now a Nobel laureate) gave a talk at an annual meeting of the American Physical Society entitled "There's Plenty of Room at the Bottom." He described a non-biochemical approach to nanomachinery (working down, step by step, using larger machines to build smaller machines), and stated that the principles of physics "do not speak against the possibility of maneuvering things atom by atom. It is not an attempt to violate any laws; it is something, in principle, that can be done; but, in practice, it has not been done because we are too big. . . . Ultimately, we can do chemical synthesis. . . . put the atoms down where the chemist says, and so you make the substance." In brief, he sketched another, nonbiochemical path to the assembler. He also stated, even then, that it is "a development which I think cannot be avoided."

As I will discuss in Chapters 4 and 5, assemblers and intelligent machines will simplify many questions regarding the time and cost of technological developments. But questions of time and cost will still muddy our view of the period between the present and these breakthroughs. Richard Feynman saw in 1959 that nanomachines could direct chemical synthesis, presumably including the synthesis of DNA. Yet he could foresee neither the time nor the cost of doing so.

In fact, of course, biochemists developed techniques for making DNA without programmable nanomachines, using shortcuts based on specific chemical tricks. Winning technologies often succeed because of unobvious tricks and details. In the mid-1950s physicists could see that basic semiconductor principles made microcircuits physically possible, but foreseeing how they would be made—foreseeing the details of mask-making, resists, oxide growth, ion implantation, etching, and so forth, in all their complexity—would have been impossible. The nuances of detail and competitive advantage that select winning technologies make the technology race complex and its path unpredictable.

But does this make long-term forecasting futile? In a race toward the limits set by natural law, the finish line is predictable even if the path and the pace of the runners are not. Not human whims but the unchanging laws of nature draw the line between what is physically possible and what is not—no political act, no social movement can change the law of gravity one whit. So however futuristic they may seem, sound projections of technological possibilities are quite distinct from predictions. They rest on timeless laws of nature, not on the vagaries of events.

It is unfortunate that this insight remains rare. Without it, we stumble in a daze across the landscape of the possible, confusing mountains with mirages and discounting both. We look ahead with minds and cultures rooted in the ideas of more sluggish times, when both science and technological competition lacked their present strength and speed. We have only recently begun to evolve a tradition of technological foresight.

SCIENCE AND NATURAL LAW

Science and technology intertwine. Engineers use knowledge produced by scientists; scientists use tools produced by engineers. Scientists and engineers both work with mathematical descriptions of natural laws and test ideas with experiments. But science and technology differ radically in their basis, methods, and aims. Understanding these differences is crucial to sound foresight. Though both fields consist of evolving meme systems, they evolve under different pressures. Consider the roots of scientific knowledge.

Through most of history, people had little understanding of evolution. This left philosophers thinking that sensory evidence, through reason, must somehow imprint on the mind all human knowledge—including knowledge of natural law. But in 1737, the Scottish philosopher David Hume presented them with a nasty puzzle: he showed that observations cannot logically prove a general rule, that the Sun shining day after day proves nothing, logically, about its shining tomorrow. And indeed, someday the Sun will fail, disproving any such logic. Hume's problem appeared to destroy the idea of rational knowledge, greatly upsetting rational philosophers (including himself). They thrashed and sweated, and irrationalism gained ground. In 1945, philosopher Bertrand Russell observed that "the growth of unreason throughout the nineteenth century and what has passed of the twentieth is a natural sequel to Hume's destruction of empiricism." Hume's problem-meme had undercut the very idea of rational knowledge, at least as people had imagined it.

In recent decades, Karl Popper (perhaps the scientists' favorite philosopher of science), Thomas Kuhn, and others have recognized science as an evolutionary process. They see it not as a mechanical process by which observations somehow generate conclusions, but as a battle where ideas compete for acceptance.

All ideas, as memes, compete for acceptance, but the meme system of science is special: it has a tradition of deliberate idea mutation, and a unique immune system for controlling the mutants. The results of evolution vary with the selective pressures applied, whether among test tube RNA molecules, insects, ideas, or machines. Hardware evolved for refrigeration differs from hardware evolved for transportation, hence refrigerators make very poor cars. In general, replicators evolved for A differ from those evolved for B. Memes are no exception.

Broadly speaking, ideas can evolve to *seem* true or they can evolve to *be* true (by seeming true to people who check ideas carefully). Anthropologists and historians have described what happens when ideas evolve to *seem* true among people lacking the methods of science; the results (the evil-spirit theory of disease, the lights-on-a-dome theory of stars, and so forth) were fairly consistent worldwide. Psychologists probing people's naïve misconceptions about how objects fall have found beliefs like those that evolved into formal "scientific" systems during the Middle Ages, before the work of Galileo and Newton.

Galileo and Newton used experiments and observations to test ideas about objects and motion, beginning an era of dramatic scientific progress: Newton evolved a theory that survived every test then available. Their method of deliberate testing killed off ideas that strayed too far from the truth, including ideas that had evolved to appeal to the naïve human mind.

This trend has continued. Further variation and testing have forced the further evolution of scientific ideas, yielding some as bizarre-seeming as the varying time and curved space of relativity, or the probabilistic particle wave functions of quantum mechanics. Even biology has discarded the special life-force expected by early biologists, revealing instead elaborate systems of invisibly small molecular machines. Ideas evolved to *be* true (or close to the truth) have again and again turned out to *seem* false—or incomprehensible. The true and the true-seeming have turned out to be as different as cars and refrigerators.

Ideas in the physical sciences have evolved under several basic selection rules. First, scientists ignore ideas that lack testable consequences; they thus keep their heads from being clogged by useless parasites. Second, scientists seek replacements for ideas that have failed tests. Finally, scientists seek ideas that make the widest possible range of exact predictions. The law of gravity, for example, describes how stones fall, planets orbit, and galaxies swirl, and makes exact predictions that leave it wide open to disproof. Its breadth and precision likewise give it broad usefulness, helping engineers both to design bridges and to plan spaceflights.

The scientific community provides an environment where such memes spread, forced by competition and testing to evolve toward power and accuracy. Agreement on the importance of testing theories holds the scientific community together through fierce controversies over the theories themselves.

Inexact, limited evidence can never *prove* an exact, general theory (as Hume showed), but it can *disprove* some theories and so help scientists choose among them. Like other evolutionary processes, science creates something positive (a growing store of useful theories) through a double negative *(dis*proof of *in*correct theories). The central role of negative evidence accounts for some of the mental upset caused by science: as an engine of disproof, it can uproot cherished beliefs, leaving psychological voids that it need not refill.

In practical terms, of course, much scientific knowledge is as solid as a rock dropped on your toe. We know Earth circles the Sun (though our senses suggest otherwise) because the theory fits endless observations, and because we know why our senses are fooled. We have more than a mere theory that atoms exist: we have bonded them to form molecules, tickled light from them, seen them under microscopes (barely), and smashed them to pieces. We have more than a mere theory of evolution: we have observed mutations, observed selection, and observed evolution in the laboratory. We have found the traces of past evolution in our planet's rocks, and have observed evolution shaping our tools, our minds, and the ideas in our minds—including the idea of evolution itself. The process of science has hammered out a unified explanation of many facts, including how people and science themselves came to be.

When science finishes disproving theories, the survivors often huddle so close together that the gap between them makes no practical difference. After all, a practical difference between two surviving theories could be tested and used to disprove one of them. The differences among modern theories of gravity, for instance, are far too subtle to trouble engineers who are planning flights through the gravity fields of space. In fact, engineers plan spaceflights using Newton's disproved theory because it is simpler than Einstein's, and is accurate enough. Einstein's theory of gravity has survived all tests so far, yet there is no absolute proof for it and there never will be. His theory makes exact predictions about everything everywhere (at least about gravitational matters), but scientists can only make approximate measurements of some things somewhere. And, as Karl Popper points out, one can always invent a theory so similar to another that existing evidence cannot tell them apart.

Though media debates highlight the shaky, disputed borders of knowledge, the power of science to build agreement remains clear. Where else has agreement on so much grown so steadily and so internationally? Surely not in politics, religion, or art. Indeed, the chief rival of science is a relative: engineering, which also evolves through proposals and rigorous testing.

SCIENCE VS. TECHNOLOGY

As IBM Director of Research Ralph E. Gomory says, "The evolution of technology development is often confused with science in the public mind." This confusion muddles our efforts at foresight.

Though engineers often tread uncertain ground, they are not doomed to do so, as scientists are. They can escape the inherent risks of proposing precise, universal scientific theories. Engineers need only show that under *particular* conditions *particular* objects will perform *well enough*. A designer need know neither the exact stress in a suspension bridge cable nor the exact stress that will break it; the cable will support the bridge so long as the first remains below the second, whatever they may be.

Though measurements cannot prove precise equality, they *can* prove inequality. Engineering results can thus be solid in a way that precise scientific theories cannot. Engineering results can even survive disproof of the scientific theories supporting them, when the new theory gives similar results. The case for assemblers, for example, will survive any possible refinements in our theory of quantum mechanics and molecular bonds.

Predicting the content of new scientific knowledge is logically impossible because it makes no sense to claim to *know* already the facts you will *learn* in the future. Predicting the details of future technology, on the other hand, is merely difficult. Science aims at knowing, but engineering aims at doing; this lets engineers speak of future achievements without paradox. They can evolve their hardware in the world of mind and computation, before cutting metal or even filling in all the details of a design.

Scientists commonly recognize this difference between scientific foresight and technological foresight: they readily make *technological* predictions about science. Scientists could and did predict the quality of Voyager's pictures of Saturn's rings, for example, though not their surprising content. Indeed, they predicted the pictures' quality while the cameras were as yet mere ideas and drawings. Their calculations used well-tested principles of optics, involving no new science.

Because science aims to understand how everything works, scientific training can be a great aid in understanding specific pieces of hardware. Still, it does not automatically bring engineering expertise; designing an airliner requires much more than a knowledge of the sciences of metallurgy and aerodynamics.

Scientists are encouraged by their colleagues and their training to focus on ideas that can be tested with available apparatus. The resulting short-term focus often serves science well: it keeps scientists from wandering off into foggy worlds of untested fantasy, and swift testing makes for an efficient mental immune system. Regrettably, though, this cultural bias toward short-term testing may make scientists less interested in long-term advances in technology.

The impossibility of genuine foresight regarding science leads many scientists to regard all statements about future developments as "speculative"—a term that makes perfect sense when applied to the future of science, but little sense when applied to well-grounded projections in technology. But most engineers share similar leanings toward the short term. They too are encouraged by their training, colleagues, and employers to focus on just one kind of problem: the design of systems that can be made with present technology or with technology just around the corner. Even long-term engineering projects like the space shuttle must have a technology cutoff date after which no new developments can become part of the basic design of the system.

In brief, scientists refuse to predict future scientific knowledge, and seldom discuss future engineering developments. Engineers do project future developments, but seldom discuss any not based on present abilities. Yet this leaves a crucial gap: what of engineering developments firmly based on *present science* but awaiting *future abilities?* This gap leaves a fruitful area for study.

Imagine a line of development which involves using existing tools to build new tools, then using *those* tools to build novel hardware (perhaps including yet another generation of tools). Each set of tools may rest on established principles, yet the whole development sequence may take many years, as each step brings a host of specific problems to iron out. Scientists planning their next experiment and engineers designing their next device may well ignore all but the first step. Still, the end result may be foreseeable, lying well within the bounds of the possible shown by established science.

Recent history illustrates this pattern. Few engineers considered building space stations before rockets reached orbit, but the principles were clear enough, and space systems engineering is now a thriving field. Similarly, few mathematicians and engineers studied the possibilities of computation until computers were built, though many did afterward. So it is not too surprising that few scientists and engineers have yet examined the future of nanotechnology, however important it may become.

THE LESSON OF LEONARDO

Efforts to project engineering developments have a long history, and past examples illustrate present possibilities. For example, how did Leonardo da Vinci succeed in foreseeing so much, and why did he sometimes fail?

Leonardo lived five hundred years ago, his life spanning the discovery of the New World. He made projections in the form of drawings and inventions; each design may be seen as a projection that something much like it could be made to work. He succeeded as a mechanical engineer: he designed workable devices (some were not to be built for centuries) for excavating, metalworking, transmitting power, and other purposes. He failed as an aircraft engineer: we now know that his flying machines could never be made to work as described.

His successes at machine design are easy to understand. If parts can be made accurately enough, of a hard enough, strong enough material, then the design of slow-moving machines with levers, pulleys, and rolling bearings becomes a matter of geometry and leverage. Leonardo understood these quite well. Some of his "predictions" were long-range, but only because many years passed before people learned to make parts precise enough, hard enough, and strong enough to build (for instance) good ball bearings—their use came some three hundred years after Leonardo proposed them. Similarly, gears with superior, cycloidal teeth went unmade for almost two centuries after Leonardo drew them, and one of his chain-drive designs went unbuilt for almost three centuries.

His failures with aircraft are also easy to understand. Because Leonardo's age lacked a science of aerodynamics, he could neither calculate the forces on wings nor know the requirements for aircraft power and control.

Can people in our time hope to make projections regarding molecular machines as accurate as those Leonardo da Vinci made regarding metal machines? Can we avoid errors like those in his plans for flying machines? Leonardo's example suggests that we can. It may help to remember that Leonardo himself probably lacked confidence in his aircraft, and that his errors nonetheless held a germ of truth. He was right to believe that flying machines of some sort were possible—indeed, he could be certain of it because they already existed. Birds, bats, and bees proved the possibility of flight. Further, though there were no working examples of his ball bearings, gears, and chain drives, he could have confidence in their principles. Able minds had already built a broad foundation of knowledge about geometry and the laws of leverage. The required strength and accuracy of the parts may have caused him doubt, but not their interplay of function and motion. Leonardo could propose machines requiring better parts than any then known, and still have a measure of confidence in his designs.

Proposed molecular technologies likewise rest on a broad foundation of knowledge, not only of geometry and leverage, but of chemical bonding, statistical mechanics, and physics in general. This time, though, the problems of material properties and fabrication accuracy do not arise in any separate way. The properties of atoms and bonds *are* the material properties, and atoms come prefabricated and perfectly standardized. Thus we now seem better prepared for foresight than were people in Leonardo's time: we know more about molecules and controlled bonding than they knew about steel and precision machining. In addition, we can point to nanomachines that already exist in the cell as Leonardo could point to the machines (birds) already flying in the sky.

Projecting how second-generation nanomachines can be built by protein machines is surely easier than it was to project how precise steel machines would be built starting with the cruder machines of Leonardo's time. Learning to use crude machines to make more precise machines was bound to take time, and the methods were far from obvious. Molecular machines, in contrast, will be built from identical prefabricated atomic parts which need only be assembled. Making precise machines with crooked machines must have been harder to imagine then than molecular assembly is now. And besides, we know that molecular assembly happens all the time in nature. Again, we have firmer grounds for confidence than Leonardo did.

In Leonardo's time, people had scant knowledge of electricity and magnetism, and knew nothing of molecules and quantum mechanics. Accordingly, electric lights, radios, and computers would have baffled them. Today, however, the basic laws most important to engineering—those describing normal matter—seem well understood. As with surviving theories of gravity, the scientific engine of disproof has forced surviving theories of matter into close agreement.

Such knowledge is recent. Before this century people did not understand why solids were solid or why the Sun shone. Scientists did not understand the laws that governed matter in the ordinary world of molecules, people, planets, and stars. This is why our century has sprouted transistors and hydrogen bombs, and why molecular technology draws near. This knowledge brings new hopes and dangers, but at least it gives us the means to see ahead and to prepare.

When the basic laws of a technology are known, future possibilities can be foreseen (though with gaps, or Leonardo would have foreseen mechanical computers). Even when the basic laws are poorly known, as were the principles of aerodynamics in Leonardo's time, nature can demonstrate possibilities. Finally, when both science and nature point to a possibility, these lessons suggest that we take it to heart and plan accordingly.

THE ASSEMBLER BREAKTHROUGH

The foundations of science may evolve and shift, yet they will continue to support a steady, growing edifice of engineering knowhow. Eventually, assemblers will allow engineers to make whatever can be designed, sidestepping the traditional problems of materials and fabrication. Already, approximations and computer models allow engineers to evolve designs even in the absence of the tools required to implement them. All this will combine to permit foresight—and something more.

As nanotechnology advances, there will come a time when assemblers become an imminent prospect, backed by an earnest and well-funded development program. Their expected capabilities will have become clear.

By then, computer-aided design of molecular systems—which has already begun—will have grown common and sophisticated, spurred by advances in computer technology and the growing needs of molecular engineers. Using these design tools, engineers will be able to design second-generation nanosystems, including the second-generation assemblers needed to build them. What is more, by allowing enough margin for inaccuracies (and by preparing alternative designs), engineers will be able to design many systems that will work when first built—they will have evolved sound designs in a world of simulated molecules.

Consider the force of this situation: under development will be the greatest production tool in history, a truly general fabrication system able to make anything that can be designed—and a design system will already be in hand. Will everyone wait until assemblers appear before planning how to use them? Or will companies and countries respond to the pressures of opportunity and competition by designing nanosystems in advance, to speed the exploitation of assemblers when they first arrive?

This *design-ahead* process seems sure to occur; the only question is when it will start and how far it will go. Years of quiet design progress may well erupt into hardware with unprecedented suddenness in the wake of the assembler breakthrough. How well we design ahead —and what we design—may determine whether we survive and thrive, or whether we obliterate ourselves.

Because the assembler breakthrough will affect almost the whole of technology, foresight is an enormous task. Of the universe of possible mechanical devices, Leonardo foresaw only a few. Similarly, of the far broader universe of future technologies, modern minds can foresee only a few. A few advances, however, seem of basic importance.

Medical technology, the space frontier, advanced computers, and new social inventions all promise to play interlocking roles. But the assembler breakthrough will affect all of them, and more.

PART TWO

Profiles of the Possible

4

Engines of Abundance

If every tool, when ordered, or even of its own accord, could do the work that befits it . . . then there would be no need either of apprentices for the master workers or of slaves for the lords.

—ARISTOTLE

ON MARCH 27, 1981, CBS radio news quoted a NASA scientist as saying that engineers will be able to build self-replicating robots within twenty years, for use in space or on Earth. These machines would build copies of themselves, and the copies would be directed to make useful products. He had no doubt of their possibility, only of when they will be built. He was quite right.

Since 1951, when John von Neumann outlined the principles of self-replicating machines, scientists have generally acknowledged their possibility. In 1953 Watson and Crick described the structure of DNA, which showed how living things pass on the instructions that guide their construction. Biologists have since learned in increasing detail how the self-replicating molecular machinery of the cell works. They find that it follows the principles von Neumann had outlined. As birds prove the possibility of flight, so life in general proves the possibility of self-replication, at least by systems of molecular machines. The NASA scientist, however, had something else in mind.

CLANKING REPLICATORS

Biological replicators, such as viruses, bacteria, plants, and people, use molecular machines. Artificial replicators can use bulk technology instead. Since we have bulk technology today, engineers may use it to build replicators before molecular technology arrives.

The ancient myth of a magical life-force (coupled with the misconception that the increase of entropy means that everything in the universe must constantly run down) has spawned a meme saying that replicators must violate some natural law. This simply isn't so. Biochemists understand how cells replicate and they find no magic in them. Instead, they find machines supplied with all the materials, energy, and instructions needed to do the job. Cells *do* replicate; robots *could* replicate.

Advances in automation will lead naturally toward mechanical replicators, whether or not anyone makes them a specific goal. As competitive pressures force increased automation, the need for human labor in factories will shrink. Fujitsu Fanuc already runs the machining section in a manufacturing plant twenty-four hours a day with only nineteen workers on the floor during the day shift, and *none* on the floor during the night shift. This factory produces 250 machines a month, of which 100 are robots.

Eventually, robots could do all the robot-assembly work, assemble other equipment, make the needed parts, run the mines and generators that supply the various factories with materials and power, and so forth. Though such a network of factories spread across the landscape wouldn't resemble a pregnant robot, it would form a self-expanding, self-replicating system. The assembler breakthrough will surely arrive before the complete automation of industry, yet modern moves in this direction are moves toward a sort of gigantic, clanking replicator.

But how can such a system be maintained and repaired without human labor?

Imagine an automatic factory able to both test parts and assemble equipment. Bad parts fail the tests and are thrown out or recycled. If the factory can also take machines apart, repairs are easy: simply disassemble the faulty machines, test all their parts, replace any worn or broken parts, and reassemble them. A more efficient system would diagnose problems without testing every part, but this isn't strictly necessary.

A sprawling system of factories staffed by robots would be workable but cumbersome. Using clever design and a minimum of different parts and materials, engineers could fit a replicating system into a single box—but the box might still be huge, because it must contain equipment able to make and assemble many different parts. How many different parts? As many as it itself contains. How many different parts and materials would be needed to build a machine able to make and assemble so many different materials and parts? This is hard to estimate, but systems based on today's technology would use electronic chips. Making these alone would require too much equipment to stuff into the belly of a small replicator.

Rabbits replicate, but they require prefabricated parts such as vitamin molecules. Getting these from food lets them survive with less molecular machinery than they would need to make everything from scratch. Similarly, a mechanical replicator using prefabricated chips could be made somewhat simpler than one that made everything it needed. Its peculiar "dietary" requirements would also tie it to a wider "ecology" of machines, helping to keep it on a firm leash. Engineers in NASA-sponsored studies have proposed using such semireplicators in space, allowing space industry to expand with only a small input of sophisticated parts from Earth.

Still, since bulk-technology replicators must make and assemble their parts, they must contain both part-making and part-assembling machines. This highlights an advantage of molecular replicators: their parts are atoms, and atoms come ready-made.

MOLECULAR REPLICATORS

Cells replicate. Their machines copy their DNA, which directs their ribosomal machinery to build other machines from simpler molecules. These machines and molecules are held in a fluid-filled bag. Its membrane lets in fuel molecules and parts for more nanomachines, DNA, membrane, and so forth; it lets out spent fuel and scrapped components. A cell replicates by copying the parts inside its membrane bag, sorting them into two clumps, and then pinching the bag in two. Artificial replicators could be built to work in a similar way, but using assemblers instead of ribosomes. In this way, we could build cell-like replicators that are not limited to molecular machinery made from the soft, moist folds of protein molecules.

But engineers seem more likely to develop other approaches to replication. Evolution had no easy way to alter the fundamental pattern of the cell, and this pattern has shortcomings. In synapses, for example, the cells of the brain signal their neighbors by emptying bladders of chemical molecules. The molecules then jostle around until they bind to sensor molecules on the neighboring cell, sometimes triggering a neural impulse. A chemical synapse makes a slow switch, and neural impulses move slower than sound. With assemblers, molecular engineers will build entire computers smaller than a synapse and a millionfold faster.

Mutation and selection could no more make a synapse into a mechanical nanocomputer than a breeder could make a horse into a car. Nonetheless, engineers have built cars, and will also learn to build computers faster than brains, and replicators more capable than existing cells.

Some of these replicators will not resemble cells at all, but will instead resemble factories shrunk to cellular size. They will contain nanomachines mounted on a molecular framework and conveyor belts to move parts from machine to machine. Outside, they will have a set of assembler arms for building replicas of themselves, an atom or a section at a time.

How fast these replicators can replicate will depend on their assembly speed and their size. Imagine an advanced assembler that contains a million atoms: it can have as many as ten thousand moving parts, each containing an average of one hundred atoms—enough parts to make up a rather complex machine. In fact, the assembler itself looks like a box supporting a stubby robot arm a hundred atoms long. The box and arm contain devices that move the arm from position to position, and others that change the molecular tools at its tip.

Behind the box sits a device that reads a tape and provides mechanical signals that trigger arm motions and tool changes. In front of the arm sits an unfinished structure. Conveyors bring molecules to the assembler system. Some supply energy to motors that drive the tape reader and arm, and others supply groups of atoms for assembly. Atom by atom (or group by group), the arm moves pieces into place as directed by the tape; chemical reactions bond them to the structure on contact.

These assemblers will work fast. A fast enzyme, such as carbonic anhydrase or ketosteroid isomerase, can process almost a million molecules per second, even without conveyors and power-driven mechanisms to slap a new molecule into place as soon as an old one is released. It might seem too much to expect an assembler to grab a molecule, move it, and jam it into place in a mere millionth of a second. But small appendages can move to and fro very swiftly. A human arm can flap up and down several times per second, fingers can tap more rapidly, a fly can wave its wings fast enough to buzz, and a mosquito makes a maddening whine. Insects can wave their wings at about a thousand times the frequency of a human arm because an insect's wing is about a thousand times shorter.

An assembler arm will be about fifty million times shorter than a human arm, and so (as it turns out) it will be able to move back and forth about fifty million times more rapidly. For an assembler arm to move a mere million times per second would be like a human arm moving about once per minute: sluggish. So it seems a very reasonable goal.

The speed of replication will depend also on the total size of the system to be built. Assemblers will not replicate by themselves; they will need materials and energy, and instructions on how to use them. Ordinary chemicals can supply materials and energy, but nanomachinery must be available to process them. Bumpy polymer molecules can code information like a punched paper tape, but a reader must be available to translate the patterns of bumps into patterns of arm motion. Together, these parts form the essentials of a replicator: the tape supplies instructions for assembling a copy of the assembler, of the reader, of the other nanomachines, and of the tape itself.

A reasonable design for this sort of replicator will likely include several assembler arms and several more arms to hold and move workpieces. Each of these arms will add another million atoms or so. The other parts—tape readers, chemical processors, and so forth—may also be as complicated as assemblers. Finally, a flexible replicator system will probably include a simple computer; following the mechanical approach that I mentioned in Chapter 1, this will add roughly 100 million atoms. Altogether, these parts will total less than 150 million atoms. Assume instead a total of one *billion*, to leave a wide margin for error. Ignore the added capability of the additional assembler arms, leaving a still wider margin. Working at one million atoms per second, the system will still copy itself in one thousand seconds, or a bit over fifteen minutes—about the time a bacterium takes to replicate under good conditions.

Imagine such a replicator floating in a bottle of chemicals, making copies of itself. It builds one copy in one thousand seconds, thirty-six in ten hours. In a week, it stacks up enough copies to fill the volume of a human cell. In a century, it stacks up enough to make a respectable speck. If this were all that replicators could do, we could perhaps ignore them in safety.

Each copy, though, will build yet more copies. Thus the first replicator assembles a copy in one thousand seconds, the two replicators then build two more in the next thousand seconds, the four build another four, and the eight build another eight. At the end of ten hours, there are not thirty-six new replicators, but over 68 billion. In less than a day, they would weigh a ton; in less than two days, they would outweigh the Earth; in another four hours, they would exceed the mass of the Sun and all the planets combined—if the bottle of chemicals hadn't run dry long before.

Regular doubling means exponential growth. Replicators multiply exponentially unless restrained, as by lack of room or resources. Bacteria do it, and at about the same rate as the replicators just described. People replicate far more slowly, yet given time enough they, too, could overshoot any finite resource supply. Concern about population growth will never lose its importance. Concern about controlling rapid new replicators will soon become important indeed.

MOLECULES AND SKYSCRAPERS

Machines able to grasp and position individual atoms will be able to build almost anything by bonding the right atoms together in the right patterns, as I described at the end of Chapter 1. To be sure, building large objects one atom at a time will be slow. A fly, after all, contains about a million atoms for every second since the dinosaurs were young. Molecular machines can nonetheless build objects of substantial size—they build whales, after all.

To make large objects rapidly, a vast number of assemblers must cooperate, but replicators will produce assemblers by the ton. Indeed, with correct design, the difference between an assembler system and a replicator will lie entirely in the assembler's programming.

If a replicating assembler can copy itself in one thousand seconds, then it can be programmed to build something else its own size just as fast. Similarly, a ton of replicators can swiftly build a ton of something else—and the product will have all its billions of billions of billions of atoms in the right place, with only a minute fraction misplaced.

To see the abilities and limits of one method for assembling large objects, imagine a flat sheet covered with small assembly arms—perhaps an army of replicators reprogrammed for construction work and arrayed in orderly ranks. Conveyors and communication channels behind them supply reactive molecules, energy, and assembly instructions. If each arm occupies an area 100 atomic diameters wide, then behind each assembler will be room for conveyors and channels totaling about 10,000 atoms in cross-sectional area.

This seems room enough. A space ten or twenty atoms wide can hold a conveyor (perhaps based on molecular belts and pulleys). A channel a few atoms wide can hold a molecular rod which, like those in the mechanical computer mentioned in Chapter 1, will be pushed and pulled to transmit signals. All the arms will work together to build a broad, solid structure layer by layer. Each arm will be responsible for its own area, handling about 10,000 atoms per layer. A sheet of assemblers handling 1,000,000 atoms per second per arm will complete about one hundred atomic layers per second. This may sound fast, but at this rate piling up a paper-sheet thickness will take about an hour, and making a meter-thick slab will take over a year.

Faster arms might raise the assembly speed to over a meter per day, but they would produce more waste heat. If they could build a meter-thick layer in a day, the heat from one square meter could cook hundreds of steaks simultaneously, and might fry the machinery. At some size and speed, cooling problems will become a limiting factor, but there are other ways of assembling objects faster without overheating the machinery.

Imagine trying to build a house by gluing together individual grains of sand. Adding a layer of grains might take grain-gluing machines so long that raising the walls would take decades. Now imagine that machines in a factory first glue the grains together to make bricks. The factory can work on many bricks at once. With enough grain-gluing machines, bricks would pour out fast; wall assemblers could then build walls swiftly by stacking the preassembled bricks. Similarly, molecular assemblers will team up with larger assemblers to build big things quickly—machines can be any size from molecular to gigantic. With this approach, most of the assembly heat will be dissipated far from the work site, in making the parts.

Skyscraper construction and the architecture of life suggest a related way to construct large objects. Large plants and animals have vascular systems, intricate channels that carry materials to molecular machinery working throughout their tissues. Similarly, after riggers and riveters finish the frame of a skyscraper, the building's "vascular system"—its elevators and corridors, aided by cranes—carry construction materials to workers throughout the interior. Assembly systems could also employ this strategy, first putting up a scaffold and then working throughout its volume, incorporating materials brought through channels from the outside.

Imagine this approach being used to "grow" a large rocket engine, working inside a vat in an industrial plant. The vat—made of shiny steel, with a glass window for the benefit of visitors—stands taller than a person, since it must hold the completed engine. Pipes and pumps link it to other equipment and to water-cooled heat exchangers. This arrangement lets the operator circulate various fluids through the vat.

To begin the process, the operator swings back the top of the vat and lowers into it a base plate on which the engine will be built. The top is then resealed. At the touch of a button, pumps flood the chamber with a thick, milky fluid which submerges the plate and then obscures the window. This fluid flows from another vat in which replicating assemblers have been raised and then reprogrammed by making them copy and spread a new instruction tape (a bit like infecting bacteria with a virus). These new assembler systems, smaller than bacteria, scatter light and make the fluid look milky. Their sheer abundance makes it viscous.

At the center of the base plate, deep in the swirling, assembler-laden fluid, sits a "seed." It contains a nanocomputer with stored engine plans, and its surface sports patches to which assemblers stick. When an assembler sticks to it, they plug themselves together and the seed computer transfers instructions to the assembler computer. This new programming tells it where it is in relation to the seed, and directs it to extend its manipulator arms to snag more assemblers. These then plug in and are similarly programmed. Obeying these instructions from the seed (which spread through the expanding network of communicating assemblers) a sort of assembler-crystal grows from the chaos of the liquid. Since each asembler knows its location in the plan, it snags more assemblers only where more are needed. This forms a pattern less regular and more complex than that of any natural crystal. In the course of a few hours, the assembler scaffolding grows to match the final shape of the planned rocket engine.

Then the vat's pumps return to life, replacing the milky fluid of unattached assemblers with a clear mixture of organic solvents and dissolved substances—including aluminum compounds, oxygen-rich compounds, and compounds to serve as assembler fuel. As the fluid clears, the shape of the rocket engine grows visible through the window, looking like a full-scale model sculpted in translucent white plastic. Next, a message spreading from the seed directs designated assemblers to release their neighbors and fold their arms. They wash out of the structure in sudden streamers of white, leaving a spongy lattice of attached assemblers, now with room enough to work. The engine shape in the vat grows almost transparent, with a hint of iridescence.

Each remaining assembler, though still linked to its neighbors, is now surrounded by tiny fluid-filled channels. Special arms on the assemblers work like flagella, whipping the fluid along to circulate it through the channels. These motions, like all the others performed by the assemblers, are powered by molecular engines fueled by molecules in the fluid. As dissolved sugar powers yeast, so these dissolved chemicals power assemblers. The flowing fluid brings fresh fuel and dissolved raw materials for construction; as it flows out it carries off waste heat. The communications network spreads instructions to each assembler.

The assemblers are now ready to start construction. They are to build a rocket engine, consisting mostly of pipes and pumps. This means building strong, light structures in intricate shapes, some able to stand intense heat, some full of tubes to carry cooling fluid. Where great strength is needed, the assemblers set to work constructing rods of interlocked fibers of carbon, in its diamond form. From these, they build a lattice tailored to stand up to the expected pattern of stress. Where resistance to heat and corrosion is essential (as on many surfaces), they build similar structures of aluminum oxide, in its sapphire form. In places where stress will be low, the assemblers save mass by leaving wider spaces in the lattice. In places where stress will be high, the assemblers reinforce the structure until the remaining passages are barely wide enough for the assemblers to move. Elsewhere the assemblers lay down other materials to make sensors, computers, motors, solenoids, and whatever else is needed.

To finish their jobs, they build walls to divide the remaining channel spaces into almost sealed cells, then withdraw to the last openings and pump out the fluid inside. Sealing the empty cells, they withdraw completely and float away in the circulating fluid. Finally, the vat drains, a spray rinses the engine, the lid lifts, and the finished engine is hoisted out to dry. Its creation has required less than a day and almost no human attention.

What is the engine like? Rather than being a massive piece of welded and bolted metal, it is a seamless thing, gemlike. Its empty internal cells, patterned in arrays about a wavelength of light apart, have a side effect: like the pits on a laser disk they diffract light, producing a varied iridescence like that of a fire opal. These empty spaces lighten a structure already made from some of the lightest, strongest materials known. Compared to a modern metal engine, this advanced engine has over 90 percent less mass.

Tap it, and it rings like a bell of surprisingly high pitch for its size. Mounted in a spacecraft of similar construction, it flies from a runway to space and back again with ease. It stands long, hard use because its strong materials have let designers include large safety margins. Because assemblers have let designers pattern its structure to yield before breaking (blunting cracks and halting their spread), the engine is not only strong but tough.

For all its excellence, this engine is fundamentally quite conventional. It has merely replaced dense metal with carefully tailored structures of light, tightly bonded atoms. The final product contains no nanomachinery.

More advanced designs will exploit nanotechnology more deeply. They could leave a vascular system in place to supply assembler and disassembler systems; these can be programmed to mend worn parts. So long as users supply such an engine with energy and raw materials, it will renew its own structure. More advanced engines can also be literally more flexible. Rocket engines work best if they can take different shapes under different operating conditions, but engineers cannot make bulk metal strong, light, and limber. With nanotechnology, though, a structure stronger than steel and lighter than wood could change shape like muscle (working, like muscle, on the sliding-fiber principle). An engine could then expand, contract, and bend at the base to provide the desired thrust in the desired direction under varying conditions. With properly programmed assemblers and disassemblers, it could even remodel its fundamental structure long after leaving the vat.

In short, replicating assemblers will copy themselves by the ton, then make other products such as computers, rocket engines, chairs, and so forth. They will make disassemblers able to break down rock to supply raw material. They will make solar collectors to supply energy. Though tiny, they will build big. Teams of nanomachines in nature build whales, and seeds replicate machinery and organize atoms into vast structures of cellulose, building redwood trees. There is nothing too startling about growing a rocket engine in a specially prepared vat. Indeed, foresters given suitable assembler "seeds" could grow spaceships from soil, air, and sunlight.

Assemblers will be able to make virtually anything from common materials without labor, replacing smoking factories with systems as clean as forests. They will transform technology and the economy at their roots, opening a new world of possibilities. They will indeed be engines of abundance.

5

Thinking Machines

The world stands on the threshold of a second computer age. New technology now moving out of the laboratory is starting to change the computer from a fantastically fast calculating machine to a device that mimics human thought processes—giving machines the capability to reason, make judgments, and even learn. Already this "artificial intelligence" is performing tasks once thought to require human intelligence . . .

—Business Week

COMPUTERS have emerged from back rooms and laboratories to help with writing, calculating, and play in homes and offices. These machines do simple, repetitive tasks, but machines still in the laboratory do much more. Artificial intelligence researchers say that computers can be made smart, and fewer and fewer people disagree. To understand our future, we must see whether artificial intelligence is as impossible as flying to the Moon.

Thinking machines need not resemble human beings in shape, purpose, or mental skills. Indeed, some artificial intelligence systems will show few traits of the intelligent liberal arts graduate, but will instead serve only as powerful engines of design. Nonetheless, understanding how human minds evolved from mindless matter will shed light on how machines can be made to think. Minds, like other forms of order, evolved through variation and selection.

Minds act. One need not embrace Skinnerian behaviorism to see the importance of behavior, including the internal behavior called thinking. RNA replicating in test tubes shows how the idea of purpose can apply (as a kind of shorthand) to utterly mindless molecules. They lack nerves and muscles, but they have evolved to "behave" in ways that promote their replication. Variation and selection have shaped each molecule's simple behavior, which remains fixed for its whole "life."

Individual RNA molecules don't adapt, but bacteria do. Competition has favored bacteria that adapt to change, for example by adjusting their mix of digestive enzymes to suit the food available. Yet these mechanisms of adaptation are themselves fixed: food molecules trip genetic switches as cold air trips a thermostat.

Some bacteria also use a primitive form of trial-and-error guidance. Bacteria of this sort tend to swim in straight lines, and have just enough "memory" to know whether conditions are improving or worsening as they go. If they sense that conditions are improving, they keep going straight. If they sense that conditions are getting worse, they stop, tumble, and head off in a random, generally different, direction. They test directions, and favor the good directions by discarding the bad. And because this makes them wander toward concentrations of food molecules, they have prospered.

Flatworms lack brains, yet show the faculty of true learning. They can learn to choose the correct path in a simple T-maze. They try turning left and turning right, and gradually select the behavior—or form the habit—which produces the better result. This is selection of behavior by its consequences, which behaviorist psychologists call "the Law of Effect." The evolving genes of worm species have produced worm individuals with evolving behavior.

Still, worms trained to run mazes (even Skinner's pigeons, trained to peck when a light flashes green) show no sign of the reflective thought we associate with mind. Organisms adapting only though the simple Law of Effect learn only by trial and error, by varying and selecting actual behavior—they don't think ahead and decide. Yet natural selection often favored organisms that could think, and thinking is not magical. As Daniel Dennett of Tufts University points out, evolved genes can equip animal brains with internal models of how the world works (somewhat like the models in computer-aided engineering systems). The animals can then "imagine" various actions and consequences, avoiding actions which "seem" dangerous and carrying out actions which "seem" safe and profitable. By testing ideas against these internal models, they can save the effort and risk of testing actions in the external world.

Dennett further points out that the Law of Effect can reshape the models themselves. As genes can provide for evolving behavior, so they can provide for evolving mental models. Flexible organisms can vary their models and pay more attention to the versions that prove better guides to action. We all know what it is to try things, and learn which work. Models need not be instinctive; they can evolve in the course of a single life.

Speechless animals, however, seldom pass on their new insights. These vanish with the brain that first produced them, because learned mental models are not stamped into the genes. Yet even speechless animals can imitate each other, giving rise to memes and cultures. A female monkey in Japan invented a way to use water to separate grain from sand; others quickly learned to do the same. In human cultures, with their language and pictures, valuable new models of how the world works can outlast their creators and spread worldwide.

On a still higher level, a mind (and "mind" is by now a fitting name) can hold evolving standards for judging whether the parts of a model—the ideas of a worldview—seem reliable enough to guide action. The mind thus selects its own contents, including its selection rules. The rules of judgment that filter the contents of science evolved in this way.

As behavior, models, and standards for knowledge evolve, so can goals. That which *brings* good, as judged by some more basic standard, eventually begins to *seem* good: it then becomes a goal in itself. Honesty pays, and becomes a valued principle of action. As thought and mental models guide action and further thought, we adopt clear thinking and accurate models as goals in themselves. Curiosity grows, and with it a love of knowledge for its own sake. The evolution of goals thus brings forth both science and ethics. As Charles Darwin wrote, "the highest possible stage in moral culture is when we recognize that we ought to control our thoughts." We achieve this as well by variation and selection, by concentrating on thoughts of value and letting others slip from attention.

Marvin Minsky of the MIT Artificial Intelligence Laboratory views the mind as a sort of society, an evolving system of communicating, cooperating, competing agencies, each made up of yet simpler agents. He describes thinking and action in terms of the activity of these agencies. Some agencies can do little more than guide a hand to grasp a cup; others (vastly more elaborate) guide the speech system as it chooses words in a sticky situation. We aren't aware of directing our fingers to wrap around a cup *just so*. We delegate such tasks to competent agents and seldom notice unless they slip. We all feel conflicting impulses and speak unintended words; these are symptoms of discord among the agents of the mind. Our awareness of this is part of the self-regulating process by which our most general agencies manage the rest.

Memes may be seen as agents in the mind that are formed by teaching and imitation. To feel that two ideas conflict, you must have embodied both of them as agents in your mind—though one may be old, strong, and supported by allies, and the other a fresh idea-agent that may not survive its first battle. Because of our superficial self-awareness, we often wonder where an idea in our heads came from. Some people imagine that these thoughts and feelings come directly from agencies outside their own minds; they incline toward a belief in haunted heads.

In ancient Rome, people believed in "genii," in good and evil spirits attending a person from cradle to grave, bringing good and ill luck. They attributed outstanding success to a special "genius." And even now, people who fail to see how natural processes create novelty see "genius" as a form of magic. But in fact, evolving genes have made minds that expand their knowledge by varying idea patterns and selecting among them. With quick variation and effective selection, guided by knowledge borrowed from others, why shouldn't such minds show what we call genius? Seeing intelligence as a natural process makes the idea of intelligent machines less startling. It also suggests how they might work.

MACHINE INTELLIGENCE

One dictionary definition of "machine" is "Any system or device, such as an electronic computer, that performs or assists in the peformance of a human task." But just how many human tasks will machines be able to perform? Calculation was once a mental skill beyond machines, the province of the intelligent and educated. Today, no one thinks of calling a pocket calculator an artificial intelligence; calculation now seems a "merely" mechanical procedure.

Still, the idea of building ordinary computers once was shocking. By the mid 1800s, though, Charles Babbage had built mechanical calculators and part of a programmable mechanical computer; however, he ran into difficulties of finance and construction. One Dr. Young helped not at all: he argued that it would be cheaper to invest the money and use the interest to pay human calculators. Nor did the British Astronomer Royal, Sir George Airy—an entry in his diary states that "On September 15th Mr. Goulburn . . . asked my opinion on the utility of Babbage's calculating machine . . . I replied, entering fully into the matter, and giving my opinion that it was worthless."

Babbage's machine was ahead of its time—meaning that in building it, machinists were forced to advance the art of making precision parts. And in fact it would *not* have greatly exceeded the speed of a skilled human calculator—but it would have been more reliable and easier to improve.

The story of computers and artificial intelligence (known as AI) resembles that of flight in air and space. Until recently people dismissed both ideas as impossible—commonly meaning that they couldn't see how to do them, or would be upset if they could. And so far, AI has had no simple, clinching demonstration, no equivalent of a working airplane or a landing on the Moon. It has come a long way, but people keep changing their definitions of intelligence.

Press reports of "giant electronic brains" aside, few people called the first computers intelligent. Indeed, the very name "computer" suggests a mere arithmetic machine. Yet in 1956, at Dartmouth, during the world's first conference on artificial intelligence, researchers Alan Newell and Herbert Simon unveiled Logic Theorist, a program that proved theorems in symbolic logic. In later years computer programs were playing chess and helping chemists determine molecular structures. Two medical programs, CASNET and MYCIN (the first dealing with internal medicine, the other with the diagnosis and treatment of infections), have performed impressively. According to the *Handbook of Artificial Intelligence,* they have been "rated, in experimental evaluations, as performing at human-expert levels in their respective domains." A program called PROSPECTOR has located, in Washington state, a molybdenum deposit worth millions of dollars. These so-called "expert systems" succeed only within strictly limited areas of competence, but they would have amazed the computer programmers of the early 1950s. Today, however, few people consider them to be *real* artificial intelligence: AI has been a moving target. The passage from *Business Week* quoted earlier only shows that computers can now be programmed with enough knowledge, and perform fancy enough tricks, that some people feel comfortable calling them intelligent. Years of seeing fictional robots and talking computers on television have at least made the idea of AI familiar.

The chief reason for declaring AI impossible has always been the notion that "machines" are intrinsically stupid, an idea that is now beginning to fade. Past machines have indeed been gross, clumsy things that did simple, brute-force work. But computers handle information, follow complex instructions, and can be instructed to change their own instructions. They can experiment and learn. They contain not gears and grease but traceries of wire and evanescent patterns of electrical energy. As Douglas Hofstadter urges (through a character in a dialogue about AI), "Why don't you let the word 'machine' conjure up images of patterns of dancing light rather than of giant steam shovels?"

Cocktail-party critics confronted with the idea of artificial intelligence often point to the stupidity of present computers, as if this proved something about the future. (A future machine may wonder whether such critics exhibited genuine thought.) Their objection is irrelevant—steam locomotives didn't fly, though they demonstrated mechanical principles later used in airplane engines. Likewise, the creeping worms of an eon ago showed no noticeable intelligence, yet our brains use neurons much like theirs.

Casual critics also avoid thinking seriously about AI by declaring that we can't possibly build machines smarter than ourselves. They forget what history shows. Our distant, speechless ancestors managed to bring forth entities of greater intelligence through genetic evolution without even thinking about it. But we *are* thinking about it, and the memes of technology evolve far more swiftly than the genes of biology. We can surely make machines with a more human-like ability to learn and organize knowledge.

There seems to be only one idea that could argue for the impossibility of making thought patterns dance in new forms of matter. This is the idea of *mental materialism*—the concept that mind is a special substance, a magical thinking-stuff somehow beyond imitation, duplication, or technological use.

Psychobiologists see no evidence for such a substance, and find no need for mental materialism to explain the mind. Because the complexity of the brain lies beyond the full grasp of human understanding, it seems complex enough to embody a mind. Indeed, if a single person could *fully* understand a brain, this would make the brain less complex than that person's mind. If all Earth's billions of people could cooperate in simply watching the activity of one human brain, each person would have to monitor tens of thousands of active synapses simultaneously—clearly an impossible task. For a person to try to understand the flickering patterns of the brain as a whole would be five billion times more absurd. Since our brain's mechanism so massively overwhelms our mind's ability to grasp it, that mechanism seems complex enough to embody the mind itself.

TURING'S TARGET

In a 1950 paper on machine intelligence, British mathematician Alan Turing wrote: "I believe that by the end of the century the use of words and general educated opinion will have altered so much that one will be able to speak of machines thinking without expecting to be contradicted." But this will depend on what we call thinking. Some say that only people can think, and that computers cannot be people; they then sit back and look smug.

But in his paper, Turing asked how we judge *human* intelligence, and suggested that we commonly judge people by the quality of their conversation. He then proposed what he called the imitation game—which everyone else now calls the Turing test. Imagine that you are in a room, able to communicate through a terminal with a person and a computer in two other rooms. You type messages; both the person and the computer can reply. Each tries to act human and intelligent. After a prolonged keyboard "conversation" with them—perhaps touching on literature, art, the weather, and how a mouth tastes in the morning—it might be that you could not tell which was the person and which the machine. If a machine could converse this well on a regular basis, then Turing suggests that we should consider it genuinely intelligent. Further, we would have to acknowledge that it knew a great deal about human beings.

For most practical purposes, we need not ask "Can a machine have self-awareness—that is, consciousness?" Indeed, critics who declare that machines cannot be conscious never seem able to say quite what they mean by the term. Self-awareness evolved to guide thought and action, not merely to ornament our humanity. We must be aware of other people, and of their abilities and inclinations, to make plans that involve them. Likewise we must be aware of ourselves, and of our own abilities and inclinations, to make plans about ourselves. There is no special mystery in *self*-awareness. What we call the self reacts to impressions from the rest of the mind, orchestrating some of its activities; this makes it no more (and no less) than a special part of the interacting patterns of thought. The idea that the self is a pattern in a special mind substance (distinct from the mind substance of the brain) would explain nothing about awareness.

A machine attempting to pass the Turing test would, of course, claim to have self-awareness. Hard-core biochauvinists would simply say that it was lying or confused. So long as they refuse to say what they mean by consciousness, they can never be proved wrong. Nonetheless, whether called conscious or not, intelligent machines will still *act* intelligent, and it is their actions that will affect us. Perhaps they will someday shame the biochauvinists into silence by impassioned argument, aided by a brilliant public-relations campaign.

No machine can now pass the Turing test, and none is likely to do so soon. It seems wise to ask whether there is a good reason even to try: we may gain more from AI research guided by other goals.

Let us distinguish two sorts of artificial intelligence, though a system could show both kinds. The first is *technical AI*, adapted to deal with the physical world. Efforts in this field lead toward automated engineering and scientific inquiry. The second is *social AI*, adapted to deal with human minds. Efforts in this field lead toward machines able to pass the Turing test.

Researchers working on social AI systems will learn much about the human mind along the way, and their systems will doubtless have great practical value, since we all can profit from intelligent help and advice. But automated engineering based on technical AI will have a greater impact on the technology race, including the race toward molecular technology. And an advanced automated engineering system may be easier to develop than a Turing-test passer, which must not only possess knowledge and intelligence, but must mimic *human* knowledge and *human* intelligence—a special, more difficult challenge.

As Turing asked, "May not machines carry out something which ought to be described as thinking but which is very different from what a man does?" Although some writers and politicians may refuse to recognize machine intelligence until they are confronted with a talkative machine able to pass the Turing test, many engineers will recognize intelligence in other forms.

ENGINES OF DESIGN

We are well on the way to automated engineering. Knowledge engineers have marketed expert systems that help people to deal with practical problems. Programmers have created computer-aided design systems that embody knowledge about shapes and motion, stress and strain, electronic circuits, heat flow, and how machine tools shape metal. Designers use these systems to augment their mental models, speeding the evolution of yet unbuilt designs. Together, designers and computers form intelligent, semiartificial systems.

Engineers can use a wide variety of computer systems to aid their work. At one end of the spectrum, they use computer screens simply as drawing boards. Farther along, they use systems able to describe parts in three dimensions and calculate their response to heat, stress, current, and so on. Some systems also know about computer-controlled manufacturing equipment, letting engineers make simulated tests of instructions that will later direct computer-controlled machines to make real parts. But the far end of the spectrum of systems involves using computers not just to record and test designs, but to generate them.

Programmers have developed their most impressive tools for use in the computer business itself. Software for chip design is an example. Integrated circuit chips now contain many thousands of transistors and wires. Designers once had to work for many months to design a circuit to do a given job, and to lay out its many parts across the surface of the chip. Today they can often delegate this task to a so-called "silicon compiler." Given a specification of a chip's function, these software systems can produce a detailed design—ready for manufacture—with little or no human help.

All these systems rely entirely on human knowledge, laboriously gathered and coded. The most flexible automated design systems today can fiddle with a proposed design to seek improvements, but they learn nothing applicable to the *next* design. But EURISKO is different. Developed by Professor Douglas Lenat and others at Stanford University, EURISKO is designed to explore new areas of knowledge. It is guided by heuristics—pieces of knowledge that suggest plausible actions to follow or implausible ones to avoid; in effect, various rules of thumb. It uses heuristics to suggest topics to work on, and further heuristics to suggest what approaches to try and how to judge the results. Other heuristics look for patterns in results, propose new heuristics, and rate the value of both new and old heuristics. In this way EURISKO evolves better behaviors, better internal models, and better rules for selecting among internal models. Lenat himself describes the variation and selection of heuristics and concepts in the system in terms of "mutation" and "selection," and suggests a social, cultural metaphor for understanding their interaction.

Since heuristics evolve and compete in EURISKO, it makes sense to expect parasites to appear—as indeed many have. One machine-generated heuristic, for example, rose to the highest possible value rating by claiming to have been a co-discoverer of every valuable new conjecture. Professor Lenat has worked closely with EURISKO, improving its mental immune system by giving it heuristics for shedding parasites and avoiding stupid lines of reasoning.

EURISKO has been used to explore elementary mathematics, programming, biological evolution, games, three-dimensional integrated circuit design, oil spill cleanup, plumbing, and (of course) heuristics. In some fields it has startled its designers with novel ideas, including new electronic devices for the emerging technology of three-dimensional integrated circuits.

The results of a tournament illustrate the power of a human/AI team. Traveller TCS is a futuristic naval war game, played in accordance with two hundred pages of rules specifying design, cost, and performance constraints for the fleet ("TCS" stands for "Trillion Credit Squadron"). Professor Lenat gave EURISKO these rules, a set of starting heuristics, and a program to simulate a battle between two fleets. He reports that "it then designed fleet after fleet, using the simulator as the 'natural selection' mechanism as it 'evolved' better and better fleet designs." The program would run all night, designing, testing, and drawing lessons from the results. In the morning Lenat would cull the designs and help it along. He credits about 60 percent of the results to himself, and about 40 percent to EURISKO.

Lenat and EURISKO entered the 1981 national Traveller TCS tournament with a strange-looking fleet. The other contestants laughed at it, then lost to it. The Lenat/EURISKO fleet won every round, emerging as the national champion. As Lenat notes, "This win is made more significant by the fact that no one connected with the program had ever played this game before the tournament, or seen it played, and there were no practice rounds."

In 1982 the competition sponsors changed the rules. Lenat and EURISKO entered a very different fleet. Other contestants again laughed at it, then lost. Lenat and EURISKO again won the national championship.

In 1983 the competition sponsors told Lenat that if he entered and won again, the competition would be canceled. Lenat bowed out.

EURISKO and other AI programs show that computers need not be limited to boring, repetitive work if they are given the right sort of programming. They can explore possibilities and turn up novel ideas that surprise their creators. EURISKO has shortcomings, yet it points the way to a style of partnership in which an AI system and a human expert both contribute knowledge and creativity to a design process.

In coming years, similar systems will transform engineering. Engineers will work in a creative partnership with their machines, using software derived from current computer-aided design systems for doing simulations, and using evolving, EURISKO-like systems to suggest designs to simulate. The engineer will sit at a screen to type in goals for the design process and draw sketches of proposed designs. The system will respond by refining the designs, testing them, and displaying proposed alternatives, with explanations, graphs, and diagrams. The engineer will then make further suggestions and changes, or respond with a new task, until an entire system of hardware has been designed and simulated.

As such automated engineering systems improve, they will do more and more of the work faster and faster. More and more often, the engineer will simply propose goals and then sort among good solutions proposed by the machine. Less and less often will the engineer have to select parts, materials, and configurations. Gradually engineers will be able to propose more general goals and expect good solutions to appear as a matter of course. Just as EURISKO ran for hours evolving fleets with a Traveller TCS simulator, automated engineering systems will someday work steadily to evolve passenger jets having maximum safety and economy—or to evolve military jets and missiles best able to control the skies.

Just as EURISKO has invented electronic devices, future automated engineering systems will invent molecular machines and molecular electronic devices, aided by software for molecular simulations. Such advances in automated engineering will magnify the design-ahead phenomenon described earlier. Thus automated engineering will not only speed the assembler breakthrough, it will increase the leap that follows.

Eventually software systems will be able to create bold new designs without human help. Will most people call such systems intelligent? It doesn't really matter.

THE AI RACE

Companies and governments worldwide support AI work because it promises commercial and military advantages. The United States has many university artificial intelligence laboratories and a host of new companies with names like Machine Intelligence Corporation, Thinking Machines Corporation, Teknowledge, and Cognitive Systems Incorporated. In October of 1981 the Japanese Ministry of Trade and Industry announced a ten-year, $850 million program to develop advanced AI hardware and software. With this, Japanese researchers plan to develop systems able to perform a billion logical inferences per second. In the fall of 1984 the Moscow Academy of Science announced a similar, five-year, $100 million effort. In October of 1983 the U.S. Department of Defense announced a five-year, $600 million Strategic Computing Program; they seek machines able to see, reason, understand speech, and help manage battles. As Paul Wallich reports in the *IEEE Spectrum*, "Artificial intelligence is considered by most people to be a cornerstone of next-generation computer technology; all the efforts in different countries accord it a prominent place in their list of goals."

Advanced AI will emerge step by step, and each step will pay off in knowledge and increased ability. As with molecular technology (and many other technologies), attempts to stop advances in one city, county, or country will at most let others take the lead. A miraculous success in stopping visible AI work *everywhere* would at most delay it and, as computers grow cheaper, let it mature in secret, beyond public scrutiny. Only a world state of immense power and stability could truly stop AI research everywhere and forever—a "solution" of bloodcurdling danger, in light of past abuses of merely national power. Advanced AI seems inevitable. If we hope to form a realistic view of the future, we cannot ignore it.

In a sense, artificial intelligence will be the ultimate tool because it will help us build all possible tools. Advanced AI systems could maneuver people out of existence, or they could help us build a new and better world. Aggressors could use them for conquest, or foresighted defenders could use them to stabilize peace. They could even help us control AI itself. The hand that rocks the AI cradle may well rule the world.

As with assemblers, we will need foresight and careful strategy to use this new technology safely and well. The issues are complex and interwoven with everything from the details of molecular technology to employment and the economy to the philosophical basis of human rights. The most basic issues, though, involve what AI can do.

ARE WE SMART ENOUGH?

Despite the example of the evolution of human beings, critics may still argue that our limited intelligence may somehow prevent us from programming genuinely intelligent machines. This argument seems weak, amounting to little more than a claim that because the *critic* can't see how to succeed, no one else will ever do better. Still, few would deny that programming computers to equal human abilities will indeed require fresh insights into human psychology. Though the programming path to AI *seems* open, our knowledge does not justify the sort of solid confidence that thoughtful engineers had (decades before Sputnik) in being able to reach the Moon with rockets, or that we have today in being able to build assemblers through protein design. Programming genuine artificial intelligence, though a form of engineering, will require new science. This places it beyond firm projection.

We need accurate foresight, though. People clinging to comforting doubts about AI seem likely to suffer from radically flawed images of the future. Fortunately, automated engineering escapes some of the burden of biochauvinist prejudice. Most people are less upset by the idea of machines designing machines than they are by the idea of true general-purpose AI systems. Besides, automated engineering has been shown to work; what remains is to extend it. Still, if more general systems are likely to emerge, we would be foolish to omit them from our calculations. Is there a way to sidestep the question of our ability to design intelligent programs?

In the 1950s, many AI researchers concentrated on simulating brain functions by simulating neurons. But researchers working on programs based on words and symbols made swifter progress, and the focus of AI work shifted accordingly. Nonetheless, the basic idea of neural simulation remains sound, and molecular technology will make it more practical. What is more, this approach seems guaranteed to work because it requires no fundamental new insights into the nature of thought.

Eventually, neurobiologists will use virus-sized molecular machines to study the structure and function of the brain, cell by cell and molecule by molecule where need be. Although AI researchers may gain useful insights about the organization of thought from the resulting advances in brain science, neural simulation can succeed without such insights. Compilers translate computer programs from one language to another without understanding how they work. Photocopiers transfer patterns of words without reading them. Likewise, researchers will be able to copy the neural patterns of the brain into another medium without understanding their higher-level organization.

After learning how neurons work, engineers will be able to design and build analogous devices based on advanced nanoelectronics and nanomachines. These will interact like neurons, but will work faster. Neurons, though complex, do seem simple enough for a mind to understand and an engineer to imitate. Indeed, neurobiologists have learned much about their structure and function, even without molecular-scale machinery to probe their workings.

With this knowledge, engineers will be able to build fast, capable AI systems, even without understanding the brain and without clever programming. They need only study the brain's neural structure and join artificial neurons to form the same functional pattern. If they make all the parts right—including the way they mesh to form the whole—then the whole, too, will be right. "Neural" activity will flow in the patterns we call thought, but faster, because all the parts will work faster.

ACCELERATING THE TECHNOLOGY RACE

Advanced AI systems seem possible and inevitable, but what effect will they have? No one can answer this in full, but one effect of automated engineering is clear: it will speed our advance toward the limits of the possible.

To understand our prospects, we need some idea of how fast advanced AI systems will think. Modern computers have only a tiny fraction of the brain's complexity, yet they can already run programs imitating significant aspects of human behavior. They differ totally from the brain in their basic style of operation, though, so direct physical comparison is almost useless. The brain does a huge number of things at once, but fairly slowly; most modern computers do only one thing at a time, but with blinding speed.

Still, one can imagine AI hardware built to imitate a brain not only in function, but in structure. This might result from a neural-simulation approach, or from the evolution of AI programs to run on hardware with a brainlike style of organization. Either way, we can use analogies with the human brain to estimate a *minimum* speed for advanced assembler-built AI systems.

Neural synapses respond to signals in thousandths of a second; experimental electronic switches respond a hundred million times faster (and nanoelectronic switches will be faster yet). Neural signals travel at under one hundred meters per second; electronic signals travel a million times faster. This crude comparison of speeds suggests that brainlike electronic devices will work about a million times faster than brains made of neurons (at a rate limited by the speed of electronic signals).

This estimate *is* crude, of course. A neural synapse is more complex than a switch; it can change its response to signals by changing its structure. Over time, synapses even form and disappear. These changes in the fibers and connections of the brain embody the long-term mental changes we call learning. They have stirred Professor Robert Jastrow of Dartmouth to describe the brain as an enchanted loom, weaving and reweaving its neural patterns throughout life.

To imagine a brainlike device with comparable flexibility, picture its electronic circuits as surrounded by mechanical nanocomputers and assemblers, with one per synapse-equivalent "switch." Just as the molecular machinery of a synapse responds to patterns of neural activity by modifying the synapse's structure, so the nanocomputers will respond to patterns of activity by directing the nanomachinery to modify the switch's structure. With the right programming, and with communication among the nanocomputers to simulate chemical signals, such a device should behave almost exactly like a brain.

Despite its complexity, the device will be compact. Nanocomputers will be smaller than synapses, and assembler-built wires will be thinner than the brain's axons and dendrites. Thin wires and small switches will make for compact circuits, and compact circuits will speed the flow of electronic patterns by shortening the distances signals must travel. It seems that a structure similar to the brain will fit in less than a cubic centimeter (as discussed in the Notes). Shorter signal paths will then join with faster transmission to yield a device over ten million times faster than a human brain.

Only cooling problems might limit such machines to slower average speeds. Imagine a conservative design, a millionfold faster than a brain and dissipating a millionfold more heat. The system consists of an assembler-built block of sapphire the size of a coffee mug, honeycombed with circuit-lined cooling channels. A high-pressure water pipe of equal diameter is bolted to its top, forcing cooling water through the channels to a similar drainpipe leaving the bottom. Hefty power cables and bundles of optical-fiber data channels trail from its sides.

The cables supply fifteen megawatts of electric power. The drainpipe carries the resulting heat away in a three-ton-per-minute flow of boiling-hot water. The optical fiber bundles carry as much data as a million television channels. They bear communications with other AI systems, with engineering simulators, and with assembler systems that build designs for final testing. Every ten seconds, the system gobbles almost two kilowatt-days of electric energy (now worth about a dollar). Every ten seconds, the system completes as much design work as a human engineer working eight hours a day for a year (now worth tens of thousands of dollars). In an hour, it completes the work of centuries. For all its activity, the system works in a silence broken only by the rush of cooling water.

This addresses the question of the sheer speed of thought, but what of its complexity? AI development seems unlikely to pause at the complexity of a single human mind. As John McCarthy of Stanford's AI lab points out, if we can place the equivalent of one human mind in a metal skull, we can place the equivalent of ten thousand cooperating minds in a building. (And a large modern power plant could supply power enough for each to think at least ten thousand times as fast as a person.) To the idea of fast engineering intelligences, add the idea of fast engineering teams.

Engineering AI systems will be slowed in their work by the need to perform experiments, but not so much as one might expect. Engineers today must perform many experiments because bulk technology is unruly. Who can say in advance exactly how a new alloy will behave when forged and then bent ten million times? Tiny cracks weaken metal, but details of processing determine their nature and effects.

Because assemblers will make objects to precise specifications, the unpredictabilities of bulk technology will be avoided. Designers (whether human or AI) will then experiment only when experimentation is faster or cheaper than calculation, or (more rarely) when basic knowledge is lacking.

AI systems with access to nanomachines will perform many experiments rapidly. They will design apparatus in seconds, and replicating assemblers will build it without the many delays (ordering special parts, shipping them, and so on) that plague projects today. Experimental apparatus on the scale of an assembler, nanocomputer, or living cell will take only minutes to build, and nanomanipulators will perform a million motions per second. Running a million ordinary experiments at once will be easy. Thus, despite delays for experimentation, automated engineering systems will move technology forward with stunning speed.

From past to future, then, the likely pattern of advancing ability looks something like this. Across eons of time, life moved forward in a long, slow advance, paced by genetic evolution. Minds with language picked up the pace, accelerated by the flexibility of memes. The invention of the methods of science and technology further accelerated advances by forcing memes to evolve faster. Growing wealth, education, and population—and better physical and intellectual tools—have continued this accelerating trend across our century.

The automation of engineering will speed the pace still more. Computer-aided design will improve, helping human engineers to generate and test ideas ever more quickly. Successors to EURISKO will shrink design times by suggesting designs and filling in the details of human innovations. At some point, full-fledged automated engineering systems will pull ahead on their own.

In parallel, molecular technology will develop and mature, aided by advances in automated engineering. Then assembler-built AI systems will bring still swifter automated engineering, evolving technological ideas at a pace set by systems a million times faster than a human brain. The rate of technological advance will then quicken to a great upward leap: in a brief time, many areas of technology will advance to the limits set by natural law. In those fields, advance will then halt on a lofty plateau of achievement.

This transformation is a dizzying prospect. Beyond it, if we survive, lies a world with replicating assemblers, able to make whatever they are told to make, without need for human labor. Beyond it, if we survive, lies a world with automated engineering systems able to direct assemblers to make devices near the limits of the possible, near the final limits of technical perfection.

Eventually, some AI systems will have both great technical ability and the social ability needed to understand human speech and wishes. If given charge of energy, materials, and assemblers, such a system might aptly be called a "genie machine." What you ask for, it will produce. Arabian legend and universal common sense suggest that we take the dangers of such engines of creation very seriously indeed.

Decisive breakthroughs in technical and social AI will be years in arriving. As Marvin Minsky has said, "The modestly intelligent machines of the near future promise only to bring us the wealth and comfort of tireless, obedient, and inexpensive servants." Most systems now called "AI" do not think or learn; they are only a crude distillate of the skills of experts, preserved, packaged, and distributed for consultation.

But genuine AI will arrive. To leave it out of our expectations would be to live in a fantasy world. To expect AI is neither optimistic nor pessimistic: as always, the researcher's optimism is the technophobe's pessimism. If we do not prepare for their arrival, social AI systems could pose a grave threat: consider the damage done by the merely human intelligence of terrorists and demagogues. Likewise, technical AI systems could destabilize the world military balance, giving one side a sudden, massive lead. With proper preparation, however, artificial intelligence could help us build a future that works—for the Earth, for people, and for the advancement of intelligence in the universe. Chapter 12 will suggest an approach, as part of the more general issue of managing the transformation that assemblers and AI will bring.

Why discuss the dangers today? Because it is not too soon to start developing institutions able to deal with such questions. Technical AI is emerging today, and its every advance will speed the technology race. Artificial intelligence is but one of many powerful technologies we must learn to manage, each adding to a complex mixture of threats and opportunities.

6

The World Beyond Earth

That inverted Bowl we call The Sky
Whereunder crawling coop'd we live and die.

—*The Rubáiyát of Omar Khayyám*

THE EARTH is but a small part of the world, and the rest of the world will be important to our future. In terms of energy, materials, and room for growth, space is almost everything. In the past, successes in space have regularly fulfilled engineering projections. In the future, an open space frontier will widen the human world. Advances in AI and nanotechnology will play a crucial role.

People took ages to recognize space as a frontier. Our ancestors once saw the night sky as a black dome with tiny sparks, a light show of the gods. They couldn't imagine space travel, because they didn't even know that outer space existed.

We now know that space exists, but few people yet understand its value. This is hardly surprising. Our minds and cultures have evolved on this planet, and we have just begun to digest the idea of a frontier beyond the sky.

Only in this century did such visionary designers as Hermann Oberth and Robert Goddard show that rockets could reach space. They had confidence in this because they knew enough about fuel, engines, tankage, and structures to calculate what multistage rockets could do. Yet, in 1921 a New York *Times* editorialist chided Goddard for the notion that rockets could fly through space without air to push against, and as late as 1956 the Astronomer Royal of Britain snorted that "Space travel is utter bilge." This only showed that editorialists and astronomers were the wrong experts to ask about space hardware. In 1957, Sputnik orbited Earth, followed in 1961 by Yuri Gagarin. In 1969, the world saw footprints on the Moon.

We paid a price for ignorance, though. Because the pioneers of space technology had lacked any way to establish their case in public, they were forced to argue basic points again and again ("Yes, rockets *will* work in vacuum. . . . Yes, they really *will* reach orbit. . . .). Busy defending the basics of spaceflight, they had little time to discuss its consequences. Thus, when Sputnik startled the world and embarrassed the United States, people were unprepared: there had been no widespread debate to shape a strategy for space.

Some of the pioneers had seen what to do: build a space station and a reusable spaceship, then reach out to the Moon or asteroids for resources. But the noise of flustered politicians promptly drowned out their suggestions, and U.S. politicians clamored for a big, easy-to-understand goal. Thus was born Project Apollo, the race to land a U.S. citizen on the nearest place to plant a flag. Project Apollo bypassed building a space station and space shuttle, instead building giant missiles able to reach the Moon in one great leap. The project was glorious, it gave scientists some information, and it brought great returns through advances in technology—but at the core, it was a hollow stunt. Taxpayers saw this, congressmen saw this, and the space program shriveled.

During Apollo, old dreams held sway in the public mind, and they were simple, romantic dreams of settling other planets. Then robot instruments dissolved the dream of a jungle-clad Venus in the reality of a planet-wide oven of high-pressure poison. They erased the lines Earthbound astronomers had drawn on Mars, and with them went both canals and Martians. In their place was a Mars of craters and canyons and dry blowing dust. Sunward of Venus lay the baked rock of Mercury; starward of Mars lay rubble and ice. The planets ranged from dead to murderous, and the dream of new Earths receded to distant stars. Space seemed a dead end.

THE NEW SPACE PROGRAM

A new space program has risen from the ruin of the old. A new generation of space advocates, engineers, and entrepreneurs now aims to make space the frontier it should have been from the beginning—a place for development and use, not for empty political gestures. They have confidence in success because space development requires no breakthroughs in science or technology. Indeed, the human race could conquer space by applying the technologies of twenty years ago—and by avoiding stunt flights, we could probably do it at a profit. Space activities need not be expensive.

Consider the high cost of reaching orbit today—thousands of dollars per kilogram. Where does it come from? To a spectator at a shuttle launch, shaken by the roar and awed by the flames, the answer seems obvious: the fuel must cost a mint. Even airlines pay roughly half their direct operating costs for fuel. A rocket resembles an airliner—it is made of aluminum and stuffed with engines, controls, and electronics—but fuel makes up almost all its mass as it sits on the launch pad. Thus, one might expect fuel to account for well over half the operating cost of a rocket. But this expectation is false. In the Moon shots, the cost of the fuel needed to reach orbit amounted to less than a million dollars—a few dollars per kilogram delivered to orbit, a fraction of a percent of the total cost. Even today, fuel remains a negligible part of the cost of spaceflight.

Why is spaceflight so much costlier than air flight? In part, because spacecraft aren't made in quantity; this forces manufacturers to recover their design costs from sales of only a few units, and to make those few units by hand at great cost. Further, most spacecraft are thrown away after one use, and even shuttles are flown just a few times a year—their cost cannot be spread over several flights a day for years, as the cost of airliners can. Finally, spaceport costs are now spread over only a few flights per month, when large airports can spread their costs over many thousands. All this conspires to make each flight into space dauntingly expensive.

But studies by Boeing Aerospace Company—the people who brought inexpensive jet transportation to much of the world—show that a fleet of fully reusable shuttles, flown and maintained like airliners, would drop the cost of reaching orbit by a factor of fifty or more.

The key is not new technology, but economies of scale and changes in management style.

Space offers vast industrial opportunities. The advantages of perching observation and communications satellites on orbit are well known. Future communications satellites will be powerful enough to communicate with hand-held stations on the ground, bringing the ultimate in mobile telephone service. Companies are already moving to take advantage of zero gravity to perform delicate separation processes, to make improved pharmaceuticals; other companies plan to grow better electronic crystals. In the years before assemblers take over materials production, engineers will use the space environment to extend the abilities of bulk technology. Space industry will provide a growing market for launch services, dropping launch costs. Falling launch costs, in turn, will stimulate the growth of space industry. Rocket transportation to Earth orbit will eventually become economical.

Space planners and entrepreneurs are already looking beyond Earth orbit to the resources of the solar system. In deep space, however, rockets swiftly become too expensive for hauling freight—they gobble fuel that itself had to be hauled into space by rockets. Fuel-burning rockets are as old as Chinese fireworks, far older than "The Star-Spangled Banner." They evolved for natural reasons: compact, powerful, and useful to the military, they can punch through air and fight strong gravity. Space engineers know of alternatives, however.

Vehicles need no great blasts of power to move through the frictionless vacuum of space. Small forces can slowly and steadily push a vehicle to enormous speeds. Because energy has mass, sunlight bouncing off a thin mirror—a solar sail—provides such a force. The pull of solar gravity provides another. Together, light pressure and gravity can carry a spacecraft anywhere in the solar system and back again. Only the heat near the Sun and the drag of planetary atmospheres will limit travel, forcing sails to steer clear of them.

NASA has studied solar sails designed to be carried to space in rockets, but these must be fairly heavy and sturdy to survive the stress of launch and unfolding. Eventually, engineers will make sails in space, using a low-mass tension structure to support mirrors of thin metal film. The result will be the "lightsail," a higher-performance class of solar sail. After a year's acceleration, a lightsail can reach a speed of one hundred kilometers per second, leaving today's swiftest rockets in the dust.

If you imagine a network of graphite-fiber strands, a spinning spiderweb kilometers wide with gaps the size of football fields between the strands, you will be well on your way to imagining the structure of a lightsail. If you picture the gaps bridged by reflecting panels built of aluminum foil thinner than a soap bubble, you will have a fair idea of how it looks: many reflective panels tied close together to form a vast, rippled mosaic of mirror. Now picture a load of cargo hanging from the web like a parachutist from a parachute, while centrifugal force holds the web-slung mirror taut and flat in the void, and you almost have it.

To build lightsails with bulk technology, we must learn to make them in space; their vast reflectors will be too delicate to survive launch and unfolding. We will need to construct scaffolding structures, manufacture thin-film reflectors, and use remotely controlled robot arms in space. But space planners already aim to master construction, manufacturing, and robotics for other space applications. If we build lightsails early in the course of space development, the effort will exercise these skills without requiring the launch of much material. Though vast, the scaffolding (together with materials for many sails) will be light enough for one or two shuttle flights to lift to orbit.

A sail production facility will produce sails cheaply. The sails, once built, will be cheap to use: they will have few critical moving parts, little mass, and zero fuel consumption. They will be utterly different from rockets in form, function, and cost of operation. In fact, calculations suggest that the costs will differ by a factor of roughly a thousand, in favor of lightsails.

Today most people view the rest of the solar system as vast and inaccessible. It *is* vast; like the Earth, it will take months to circumnavigate by sail. Its apparent inaccessibility, however, has less to do with distance than with the cost of transportation via rocket.

Lightsails can smash the cost barrier, opening the door to the solar system. Lightsails will make other planets easier to reach, but this will not make planets much more useful: they will remain deadly deserts. The gravity of planets will prevent lightsails from shuttling to their surfaces, and will also handicap industry on a planet's surface. Spinning space stations can simulate gravity if it is needed, but planet-bound stations cannot escape it. Worse yet, planetary atmospheres block solar energy, spread dust, corrode metals, warm refrigerators, cool ovens, and blow things down. Even the airless Moon rotates, blocking sunlight half the time, and has gravity enough to ground lightsails beyond hope of escape. Lightsails are fast and tireless, but not strong.

The great and enduring value of space lies in its resources of matter, energy, and room. The planets occupy room and block energy. The material resources they offer are inconveniently placed. The asteroids, in contrast, are flying mountains of resources that trace orbits crisscrossing the entire solar system. Some cross the orbit of Earth; some have even struck Earth, blasting craters. Mining the asteroids seems practical. We may need roaring rockets to carry things up *into* space, but meteorites prove that ordinary rocks can fall down *from* space—and like the space shuttle, objects falling from space need not burn up on the way down. Delivering packages of material from an asteroid to a landing target in a salt flat will cost little.

Even small asteroids are big in human terms: they hold billions of tons of resources. Some asteroids contain water and a substance resembling oil shale. Some contain fairly ordinary rock. Some contain a metal that holds elements scarce in Earth's crust, elements that sank beyond reach ages ago in the formation of Earth's metal core: this meteoritic steel is a strong, tough alloy of iron, nickel, and cobalt, bearing valuable amounts of platinum-group metals and gold. A kilometer-wide chunk of this material (and there are many) contains precious metals worth several trillion dollars, mixed with enough nickel and cobalt to supply Earth's industry for many years.

The Sun floods space with easily collected energy. A square-kilometer framework holding metal-film reflectors will gather over a billion watts of sunlight, free of interference from cloud or night. In the weatherless calm of space, the flimsiest collector will be as permanent as a hydroelectric dam. Since the Sun puts out as much energy in a microsecond as the human race now uses in a year, energy need not be scarce for some time to come.

Finally, space itself offers room to live. People once saw life in space in terms of planets. They imagined domed cities built on planets, dead planets slowly converted into Earth-like planets, and Earth-like planets reached after years in a flight to the stars. But planets are package deals, generally offering the wrong gravity, atmosphere, length of day, and location.

Free space offers a better building site for settlements. Professor Gerard O'Neill of Princeton University brought this idea to public attention, helping to revive interest in space after the post-Apollo crash. He showed that ordinary construction materials—steel and glass—could be used to build habitable cylinders in space, kilometers in length and circumference. In his design, dirt underfoot shields inhabitants from the natural radiation of space, just as Earth's inhabitants are shielded by the air overhead. Rotation produces an acceleration equaling Earth's gravity, and broad mirrors and window panels flood the interior with sunlight. Add soil, streams, vegetation, and imagination, and the lands inside could rival the best valleys on Earth as places to live. With just the resources of the asteroids, we will be able to build the practical equivalent of a thousand new Earths.

By adapting present technology, we could open the space frontier. The prospect is heartening. It shows us an obvious way to bypass terrestrial limits to growth, lessening one of the fears that has clouded our view of the future. The promise of the space frontier can thus mobilize human hope—a resource we will need in abundance, if we are to deal with other problems.

SPACE AND ADVANCED TECHNOLOGY

By adapting present technology, we could indeed open the space frontier—but we won't. Along the path foreseen by the current space movement, human civilization would take decades to become firmly established in space. Before then, breakthroughs in technology will open new paths.

Nowadays, teams of engineers typically take five to ten years to develop a new space system, spending tens to thousands of millions of dollars along the way. These engineering delays and costs make progress painfully slow. In coming years, though, computer-aided design systems will evolve toward automated engineering systems. As they do, engineering delays and costs will shrink and then plummet; computer-controlled manufacturing systems will drop overall costs still further. A day will come when automated design and manufacturing will have made space systems development more than tenfold faster and cheaper. Our progress in space will soar.

At that time, will space settlers look back on our present space program as the key to space development? Perhaps not. They will have seen more technical progress made in a few years than space engineers previously managed in a few decades. They may well conclude that AI and robotics did more for space development than did a whole army of NASA engineers.

The assembler breakthrough and automated engineering will combine to bring advances that will make our present space efforts seem quaint. In Chapter 4, I described how replicating assemblers will be able to build a light, strong rocket engine using little human labor. Using similar methods, we will build entire spacecraft of low cost and extraordinary performance. Weight for weight, their diamond-based structural materials will have roughly fifty times the strength (and fourteen times the stiffness) of the aluminum used in the present shuttle; vehicles built with these materials can be made over 90 percent lighter than similar vehicles today. Once in space, vehicles will spread solar collectors to gather abundant energy. Using this energy to power assemblers and disassemblers, they will rebuild themselves in flight to suit changing conditions or the whim of their passengers. Today, space travel is a challenge. Tomorrow, it will be easy and convenient.

Since nanotechnology lends itself to making small things, consider the smallest person-carrying spacecraft: the spacesuit. Forced to use weak, heavy, passive materials, engineers now make bulky, clumsy spacesuits. A look at an advanced spacesuit will illustrate some of the capabilites of nanotechnology.

Imagine that you are aboard a space station, spun to simulate Earth's normal gravity. After instruction, you have been given a suit to try out: there it hangs on the wall, a gray, rubbery-looking thing with a transparent helmet. You take it down, heft its substantial weight, strip, and step in through the open seam on the front.

The suit feels softer than the softest rubber, but has a slick inner surface. It slips on easily and the seam seals at a touch. It provides a skintight covering like a thin leather glove around your fingers, thickening as it runs up your arm to become as thick as your hand in the region around your torso. Behind your shoulders, scarcely noticeable, is a small backpack. Around your head, almost invisible, is the helmet. Below your neck the suit's inner surface hugs your skin with a light, uniform touch that soon becomes almost imperceptible.

You stand up and walk around, experimenting. You bounce on your toes and feel no extra weight from the suit. You bend and stretch and feel no restraint, no wrinkling, no pressure points. When you rub your fingers together they feel sensitive, as if bare—but somehow slightly *thicker*. As you breathe, the air tastes clean and fresh. In fact, you feel that you could forget that you are wearing a suit at all. What is more, you feel just as comfortable when you step out into the vacuum of space.

The suit manages to do all this and more by means of complex activity within a structure having a texture almost as intricate as that of living tissue. A glove finger a millimeter thick has room for a thousand micron-thick layers of active nanomachinery and nanoelectronics. A fingertip-sized patch has room for a billion mechanical nanocomputers, with 99.9 percent of the volume left over for other components.

In particular, this leaves room for an active structure. The middle layer of the suit material holds a three-dimensional weave of diamond-based fibers acting much like artificial muscle, but able to push as well as pull (as discussed in the Notes). These fibers take up much of the volume and make the suit material as strong as steel. Powered by microscopic electric motors and controlled by nanocomputers, they give the suit material its supple strength, making it stretch, contract, and bend as needed. When the suit felt soft earlier, this was because it had been programmed to *act* soft. The suit has no difficulty holding its shape in a vacuum; it has strength enough to avoid blowing up like a balloon. Likewise, it has no difficulty supporting its own weight and moving to match your motions, quickly, smoothly, and without resistance. This is one reason why it almost seems not to be there at all.

Your fingers feel almost bare because you feel the texture of what you touch. This happens because pressure sensors cover the suit's surface and active structure covers its lining: the glove feels the shape of whatever you touch—and the detailed pattern of pressure it exerts —and transmits the same texture pattern to your skin. It also reverses the process, transmitting to the outside the detailed pattern of forces exerted by your skin on the inside of the glove. Thus the glove pretends that it isn't there, and your skin feels almost bare.

The suit has the strength of steel and the flexibility of your own body. If you reset the suit's controls, the suit continues to match your motions, but with a difference. Instead of simply transmitting the forces you exert, it amplifies them by a factor of ten. Likewise, when something brushes against you, the suit now transmits only a tenth of the force to the inside. You are now ready for a wrestling match with a gorilla.

The fresh air you breathe may not seem surprising; the backpack includes a supply of air and other consumables. Yet after a few days outside in the sunlight, your air will not run out: like a plant, the suit absorbs sunlight and the carbon dioxide you exhale, producing fresh oxygen. Also like a plant (or a whole ecosystem), it breaks down other wastes into simple molecules and reassembles them into the molecular patterns of fresh, wholesome food. In fact, the suit will keep you comfortable, breathing, and well fed almost anywhere in the inner solar system.

What is more, the suit is durable. It can tolerate the failure of numerous nanomachines because it has so many others to take over the load. The space between the active fibers leaves room enough for assemblers and disassemblers to move about and repair damaged devices. The suit repairs itself as fast as it wears out.

Within the bounds of the possible, the suit could have many other features. A speck of material smaller than a pinhead could hold the text of every book ever published, for display on a fold-out screen. Another speck could be a "seed" containing the blueprints for a range of devices greater than the total the human race has yet built, along with replicating assemblers able to make any or all of them.

What is more, fast technical AI systems like those described in the last chapter could design the suit in a morning and have it built by afternoon.

All that we accomplish in space with modern bulk technology will be swiftly and dramatically surpassed shortly after molecular technology and automated engineering arrive. In particular, we will build replicating assemblers that work in space. These replicators will use solar energy as plants do, and with it they will convert asteroidal rubble into copies of themselves and products for human use. With them, we will grasp the resources of the solar system.

By now, most readers will have noted that this, like certain earlier discussions, sounds like science fiction. Some may be pleased, some dismayed that future possibilities do in fact have this quality. Some, though, may feel that "sounding like science fiction" is somehow grounds for dismissal. This feeling is common and deserves scrutiny.

Technology and science fiction have long shared a curious relationship. In imagining future technologies, SF writers have been guided partly by science, partly by human longings, and partly by the market demand for bizarre stories. Some of their imaginings later become real, because ideas that seem plausible and interesting in fiction sometimes prove possible and attractive in actuality. What is more, when scientists and engineers foresee a dramatic possibility, such as rocket-powered spaceflight, SF writers commonly grab the idea and popularize it.

Later, when engineering advances bring these possibilities closer to realization, other writers examine the facts and describe the prospects. These descriptions, unless they are quite abstract, then sound like science fiction. Future possibilities will often resemble today's fiction, just as robots, spaceships, and computers resemble yesterday's fiction. How could it be otherwise? Dramatic new technologies sound like science fiction because science fiction authors, despite their frequent fantasies, aren't blind and have a professional interest in the area.

Science fiction authors often fictionalize (that is, counterfeit) the scientific content of their stories to "explain" dramatic technologies. Some fuzzy thinkers then take all descriptions of dramatic technical advances, lump them together with this bogus science, and ignore the lot. This is unfortunate. When engineers project future abilities, they test their ideas, evolving them to fit our best understanding of the laws of nature. The resulting concepts must be distinguished from ideas evolved to fit the demands of paperback fiction. Our lives will depend on it.

Much will remain impossible, even with molecular technology. No spacesuit, however marvelous, will be able to rocket back and forth indefinitely at tremendous speeds, or survive great explosions, or walk through walls, or even stay cool indefinitely in a hot isolated room. We have far to go before reaching the limits of the possible, yet limits exist. But this is a topic taken up later.

ABUNDANCE

Space resources join with assemblers and automated engineering systems to round out the case for a future of great material abundance. What this means can best be seen by examining costs.

Costs reflect the limits of our resources and abilities; high costs indicate scarce resources and difficult goals. The prophets of scarcity have in effect predicted steeply rising resource costs, and with them a certain kind of future. Resource costs, however, always depend on technology. Unfortunately, engineers attempting to predict the cost of future technologies have generally encountered a tangle of detail and uncertainty that proves impossible to untie. This problem has obscured our understanding of the future.

The prospect of replicating assemblers, automated engineering, and space resources cuts this Gordian knot of cost prediction. Today the cost of products includes the costs of labor, capital, raw materials, energy, land, waste disposal, organization, distribution, taxation, and design. To see how total costs will change, consider these elements one by one.

Labor. Replicating assemblers will require no labor to build, once the first exists. What use are human hands in running an assembler? Further, with robotic devices of various sizes to assemble parts into larger systems, the entire manufacturing process from assembling molecules to assembling skyscrapers could be free of labor costs.

Capital. Assembler-based systems, if properly programmed, will themselves *be* productive capital. Together with larger robotic machines, they will be able to build virtually anything, including copies of themselves. Since this self-replicating capital will be able to double many times per day, only demand and available resources will limit its quantity. Capital as such need cost virtually nothing.

Raw materials. Since molecular machines will arrange atoms to best advantage, a little material can go a long way. Common elements like hydrogen, carbon, nitrogen, oxygen, aluminum, and silicon seem best for constructing the bulk of most structures, vehicles, computers, clothes and so forth: they are light and form strong bonds. Because dirt and air contain these elements in abundance, raw materials can be dirt cheap.

Energy. Assemblers will be able to run off chemical or electrical energy. Assembler-built systems will convert solar to chemical energy, like plants, or solar to electrical energy, like solar cells. Existing solar cells are already more efficient than plants. With replicating assemblers to build solar collectors, fuel and electric power will cost little.

Land. Assembler-based production systems will occupy little room. Most could sit in a closet (or a thimble, or a pinhole); larger systems could be placed underground or in space if someone wants something that requires an unsightly amount of room. Assembler-based production systems will make both digging machines and spacecraft cheap.

Waste disposal. Assembler systems will be able to keep control of the atoms they use, making production as clean as a growing apple tree, or cleaner. If the orchard remains too dirty or ugly, we will be able to move it off Earth entirely.

Organization. Today, factory production requires organization to coordinate hordes of workers and managers. Assembler-based production machines will contain no people, and will simply sit around and produce things made to order. Their initial programming will provide all the organization and information needed to make a wide range of products.

Distribution. With automatic vehicles running in tunnels made by cheap digging machines, distribution need neither consume labor nor blight the landscape. With assemblers in the home and community, there will be less need for distribution in the first place.

Taxation. Most taxes take a fixed percentage of a price, and thus add a fixed percentage to the cost. If the cost is negligible, the tax will be negligible. Further, governments with their own replicators and raw materials will have less reason to tax people.

Design. The above points add up to a case for low costs of production. Technical AI systems, by avoiding the labor cost of engineering, will virtually eliminate the costs of design. These AI systems will themselves be inexpensive to produce and operate, being constructed by assemblers and having no inclination to do anything but design things.

In short, at the end of a long line of profitable developments in computer and molecular technologies, the cost of designing and producing things will drop dramatically. I above referred to "dirt cheap" raw materials, and indeed, assemblers will be able to make almost anything from dirt and sunlight. Space resources, however, will change "dirt cheap" to "cheap-dirt cheap": topsoil has value in Earth's ecosystem, but rubble from asteroids will come from a dead and dreary desert. By the same token, assemblers in space will run off cheap sunlight.

Space resources are vast. One asteroid could bury Earth's continents a kilometer deep in raw materials. Space swallows the 99.999999955 percent of the Sun's light that misses Earth, and most is lost to the interstellar void.

Space holds matter, energy, and room enough for projects of vast size, including vast space settlements. Replicator-based systems will be able to construct worlds of continental scale, resembling Dr. O'Neill's cylinders but made of strong, carbon-based materials. With these materials and water from the ice moons of the outer solar system, we will be able to create not only lands in space, but whole seas, wider and deeper than the Mediterranean. Constructed with energy and materials from space, these broad new lands and seas will cost Earth and its people almost nothing in terms of resources. The chief requirement will be programming the first replicator, but AI systems will help with that. The greatest problem will be deciding what we want.

As Konstantin Tsiolkovsky wrote near the turn of the century, "Man will not always stay on Earth; the pursuit of light and space will lead him to penetrate the bounds of the atmosphere, timidly at first, but in the end to conquer the whole of solar space." To dead space we will bring life.

And replicators will give us the resources to reach for the stars. A lightsail driven starward only by sunlight would soon find itself coasting in the dark—faster than any modern rocket, yet so slowly that it would take millennia to cross the interstellar gulf. We can build a tremendous bank of lasers orbiting the Sun, however, and with it drive a beam far beyond our solar system, pushing a sail toward the speed of light. The crossing then will take only years.

Stopping presents a problem. Freeman Dyson of Princeton suggests braking with magnetic fields in the thin ionized gas between the stars. Robert Forward of Hughes Research Laboratories suggests bouncing laser light off the sail, directing light back along the sail's path to decelerate a smaller sail trailing behind. One way or another (and there are many others), the stars themselves lie within our reach.

For a long time to come, however, the solar system can provide room enough. The space near Earth holds room for lands with a million times Earth's area. Nothing need stop emigration, or return visits to the old country. We will have no trouble powering the transportation system—the sunlight falling on Earth supplies enough energy in ten minutes to put today's entire population in orbit. Space travel and space settlements will both become cheap. If we make wise use of molecular technology, our descendants will wonder what kept us bottled up on Earth for so long, and in such poverty.

THE POSITIVE-SUM SOCIETY

It might seem that the cost of everything—even land, if one doesn't crave thousands of kilometers of rock underfoot—will drop to nothing. In a sense, this is almost right; in another sense, it is quite false. People will always value matter, energy, information, and genuine human service, therefore everything will still have its cost. And in the long run, we will face real limits to growth, so the cost of resources cannot be dismissed.

Nonetheless, if we survive, replicators and space resources will bring a long era in which genuine resource limits do not yet pinch us —an era when by our present standards even vast wealth will seem virtually free. This may seem too good to be true, but nature (as usual) has not set her limits based on human feelings. Our ancestors once thought that talking to someone across the sea (many months' voyage by sailing ship) would be too good to be true, but undersea cables and oversea satellites worked anyway.

But there is another, less pleasant answer for those who think assemblers are too good to be true: assemblers also threaten to bring hazards and weapons more dangerous than any yet seen. If nanotechnology could be avoided but not controlled, then sane people would shun it. The technology race, however, will bring forth assemblers from biotechnology as surely as it brought forth spacecraft from missiles. The military advantages alone will be enough to make advances almost inevitable. Assemblers are unavoidable, but perhaps controllable.

Our challenge is to avoid the dangers, but this will take cooperation, and we are more likely to cooperate if we understand how much we have to gain from it. The prospect of space and replicating assemblers may help us clear away some ancient and dangerous memes.

Human life was once like a zero-sum game. Humankind lived near its ecological limit and tribe fought tribe for living space. Where pastures, farmland, and hunting grounds were concerned, more for one group meant less for another. Because one's gain roughly equaled the other's loss, net benefits summed to zero. Still, people who cooperated on other matters prospered, and so our ancestors learned not just to grab, but to cooperate and build.

Where taxes, transfer payments, and court battles are concerned, more for one still means less for another. We add to total wealth slowly, but redistribute it swiftly. On any given day our resources seem fixed, and this gives rise to the illusion that life is a zero-sum game. This illusion suggests that broad cooperation is pointless, because our gain must result from some opponent's loss.

The history of human advance proves that the world game can be positive-sum. Accelerating economic growth during recent centuries shows that the rich can get richer while the poor get richer. Despite population growth (and the idea of dividing a fixed pie) the *average* wealth per capita worldwide, including that of the Third World, has grown steadily larger. Economic fluctuations, local reversals, and the natural tendency of the media to focus on bad news—these combine to obscure the facts about economic growth, but public records show it clearly enough. Space resources and replicating assemblers will accelerate this historic trend beyond the dreams of economists, launching the human race into a new world.

7

Engines of Healing

One of the things which distinguishes ours from all earlier generations is this, that we have seen our atoms.

—KARL K. DARROW, *The Renaissance of Physics*

WE WILL USE molecular technology to bring health because the human body is made of molecules. The ill, the old, and the injured all suffer from misarranged patterns of atoms, whether misarranged by invading viruses, passing time, or swerving cars. Devices able to rearrange atoms will be able to set them right. Nanotechnology will bring a fundamental breakthrough in medicine.

Physicians now rely chiefly on surgery and drugs to treat illness. Surgeons have advanced from stitching wounds and amputating limbs to repairing hearts and reattaching limbs. Using microscopes and fine tools, they join delicate blood vessels and nerves. Yet even the best microsurgeon cannot cut and stitch finer tissue structures. Modern scalpels and sutures are simply too coarse for repairing capillaries, cells, and molecules. Consider "delicate" surgery from a cell's perspective: a huge blade sweeps down, chopping blindly past and through the molecular machinery of a crowd of cells, slaughtering thousands. Later, a great obelisk plunges through the divided crowd, dragging a cable as wide as a freight train behind it to rope the crowd together again. From a cell's perspective, even the most delicate surgery, performed with exquisite knives and great skill, is still a butcher job. Only the ability of cells to abandon their dead, regroup, and multiply makes healing possible.

Yet as many paralyzed accident victims know too well, not all tissues heal.

Drug therapy, unlike surgery, deals with the finest structures in cells. Drug molecules are simple molecular devices. Many affect specific molecules in cells. Morphine molecules, for example, bind to certain receptor molecules in brain cells, affecting the neural impulses that signal pain. Insulin, beta blockers, and other drugs fit other receptors. But drug molecules work without direction. Once dumped into the body, they tumble and bump around in solution haphazardly until they bump a target molecule, fit, and stick, affecting its function.

Surgeons can see problems and plan actions, but they wield crude tools; drug molecules affect tissues at the molecular level, but they are too simple to sense, plan, and act. But molecular machines directed by nanocomputers will offer physicians another choice. They will combine sensors, programs, and molecular tools to form systems able to examine and repair the ultimate components of individual cells. They will bring surgical control to the molecular domain.

These advanced molecular devices will be years in arriving, but researchers motivated by medical needs are already studying molecular machines and molecular engineering. The best drugs affect specific molecular machines in specific ways. Penicillin, for example, kills certain bacteria by jamming the nanomachinery they use to build their cell walls, yet it has little effect on human cells.

Biochemists study molecular machines both to learn how to build them and to learn how to wreck them. Around the world (and especially the Third World) a disgusting variety of viruses, bacteria, protozoa, fungi, and worms parasitize human flesh. Like penicillin, safe, effective drugs for these diseases would jam the parasite's molecular machinery while leaving human molecular machinery unharmed. Dr. Seymour Cohen, professor of pharmacological science at SUNY (Stony Brook, New York), argues that biochemists should systematically study the molecular machinery of these parasites. Once biochemists have determined the shape and function of a vital protein machine, they then could often design a molecule shaped to jam it and ruin it. Such drugs could free humanity from such ancient horrors as schistosomiasis and leprosy, and from new ones such as AIDS.

Drug companies are already redesigning molecules based on knowledge of how they work. Researchers at Upjohn Company have designed and made modified molecules of vasopressin, a hormone that consists of a short chain of amino acids. Vasopressin increases the work done by the heart and decreases the rate at which the kidneys produce urine; this increases blood pressure. The researchers designed modified vasopressin molecules that affected receptor molecules in the kidney more than those in the heart, giving them more specific and controllable medical effects. More recently, they designed a modified vasopressin molecule that binds to the kidney's receptor molecules without direct effect, thus blocking and *inhibiting* the action of natural vasopressin.

Medical needs will push this work forward, encouraging researchers to take further steps toward protein design and molecular engineering. Medical, military, and economic pressures all push us in the same direction. Even before the assembler breakthrough, molecular technology will bring impressive advances in medicine; trends in biotechnology guarantee it. Still, these advances will generally be piecemeal and hard to predict, each exploiting some detail of biochemistry. Later, when we apply assemblers and technical AI systems to medicine, we will gain broader abilities that are easier to foresee.

To understand these abilities, consider cells and their self-repair mechanisms. In the cells of your body, natural radiation and noxious chemicals split molecules, producing reactive molecular fragments. These can misbond to other molecules in a process called cross-linking. As bullets and blobs of glue would damage a machine, so radiation and reactive fragments damage cells, both breaking molecular machines and gumming them up.

If your cells could not repair themselves, damage would rapidly kill them or make them run amok by damaging their control systems. But evolution has favored organisms with machinery able to do something about this problem. The self-replicating factory system sketched in Chapter 4 repaired itself by replacing damaged parts; cells do the same. So long as a cell's DNA remains intact, it can make error-free tapes that direct ribosomes to assemble new protein machines.

Unfortunately for us, DNA itself becomes damaged, resulting in mutations. Repair enzymes compensate somewhat by detecting and repairing certain kinds of damage to DNA. These repairs help cells survive, but existing repair mechanisms are too simple to correct all problems, either in DNA or elsewhere. Errors mount, contributing to the aging and death of cells—and of people.

LIFE, MIND, AND MACHINES

Does it make sense to describe cells as "machinery," whether self-repairing or not? Since we are made of cells, this might seem to reduce human beings to "mere machines," conflicting with a holistic understanding of life.

But a dictionary definition of holism is "the theory that reality is made up of organic or unified wholes that are greater than the simple sum of their parts." This certainly applies to people: one simpler sum of our parts would resemble hamburger, lacking both mind and life.

The human body includes some ten thousand billion *billion* protein parts, and no machine so complex deserves the label "mere." Any brief description of so complex a system cannot avoid being grossly incomplete, yet at the cellular level a description in terms of machinery makes sense. Molecules have simple moving parts, and many act like familiar types of machinery. Cells considered as a whole may seem less mechanical, yet biologists find it useful to describe them in terms of molecular machinery.

Biochemists have unraveled what were once the central mysteries of life, and have begun to fill in the details. They have traced how molecular machines break food molecules into their building blocks and then reassemble these parts to build and renew tissue. Many details of the structure of human cells remain unknown (single cells have billions of large molecules of thousands of different kinds), but biochemists have mapped every part of some viruses. Biochemical laboratories often sport a large wall chart showing how the chief molecular building blocks flow through bacteria. Biochemists understand much of the process of life in detail, and what they don't understand seems to operate on the same principles. The mystery of heredity has become the industry of genetic engineering. Even embryonic development and memory are being explained in terms of changes in biochemistry and cell structure.

In recent decades, the very quality of our remaining ignorance has changed. Once, biologists looked at the process of life and asked, "How can this be?" But today they understand the general principles of life, and when they study a specific living process they commonly ask, "Of the many ways this could be, which has nature chosen?" In many instances their studies have narrowed the competing explanations to a field of one. Certain biological processes—the coordination of cells to form growing embryos, learning brains, and reacting immune systems—still present a real challenge to the imagination. Yet this is not because of some deep mystery about how their parts work, but because of the immense complexity of how their many parts interact to form a whole.

Cells obey the same natural laws that describe the rest of the world. Protein machines in the right molecular environment will work whether they remain in a functioning cell or whether the rest of the cell was ground up and washed away days before. Molecular machines know nothing of "life" and "death."

Biologists—when they bother—sometimes define life as the ability to grow, replicate, and respond to stimuli. But by this standard, a mindless system of replicating factories might qualify as life, while a conscious artificial intelligence modeled on the human brain might not. Are viruses alive, or are they "merely" fancy molecular machines? No experiment can tell, because nature draws no line between living and nonliving. Biologists who work with viruses instead ask about viability: "Will this virus function, if given a chance?" The labels of "life" and "death" in medicine depend on medical capabilities: physicians ask, "Will this patient function, if we do our best?" Physicians once declared patients dead when the heart stopped; they now declare patients dead when they despair of restoring brain activity. Advances in cardiac medicine changed the definition once; advances in brain medicine will change it again.

Just as some people feel uncomfortable with the idea of machines thinking, so some feel uncomfortable with the idea that machines underlie our own thinking. The word "machine" again seems to conjure up the wrong image, a picture of gross, clanking metal, rather than signals flickering through a shifting weave of neural fibers, through a living tapestry more intricate than the mind it embodies can fully comprehend. The brain's really machinelike machines are of molecular size, smaller than the finest fibers.

A whole need not resemble its parts. A solid lump scarcely resembles a dancing fountain, yet a collection of solid, lumpy molecules forms fluid water. In a similar way, billions of molecular machines make up neural fibers and synapses, thousands of fibers and synapses make up a neural cell, billions of neural cells make up the brain, and the brain itself embodies the fluidity of thought.

To say that the mind is "just molecular machines" is like saying that the Mona Lisa is "just dabs of paint." Such statements confuse the parts with the whole, and confuse matter with the pattern it embodies. We are no less human for being made of molecules.

FROM DRUGS TO CELL REPAIR MACHINES

Being made of molecules, and having a human concern for our health, we will apply molecular machines to biomedical technology. Biologists already use antibodies to tag proteins, enzymes to cut and splice DNA, and viral syringes (like the T4 phage) to inject edited DNA into bacteria. In the future, they will use assembler-built nanomachines to probe and modify cells.

With tools like disassemblers, biologists will be able to study cell structures in ultimate, molecular detail. They then will catalog the hundreds of thousands of kinds of molecules in the body and map the structure of the hundreds of kinds of cells. Much as engineers might compile a parts list and make engineering drawings for an automobile, so biologists will describe the parts and structures of healthy tissue. By that time, they will be aided by sophisticated technical AI systems.

Physicians aim to make tissues healthy, but with drugs and surgery they can only encourage tissues to repair themselves. Molecular machines will allow more direct repairs, bringing a new era in medicine.

To repair a car, a mechanic first reaches the faulty assembly, then identifies and removes the bad parts, and finally rebuilds or replaces them. Cell repair will involve the same basic tasks—tasks that living systems already prove possible.

Access. White blood cells leave the bloodstream and move through tissue, and viruses enter cells. Biologists even poke needles into cells without killing them. These examples show that molecular machines can reach and enter cells.

Recognition. Antibodies and the tail fibers of the T4 phage—and indeed, all specific biochemical interactions—show that molecular systems can recognize other molecules by touch.

Disassembly. Digestive enzymes (and other, fiercer chemicals) show that molecular systems can disassemble damaged molecules.

Rebuilding. Replicating cells show that molecular systems can build or rebuild every molecule found in a cell.

Reassembly. Nature also shows that separated molecules can be put back together again. The machinery of the T4 phage, for example, self-assembles from solution, apparently aided by a single enzyme. Replicating cells show that molecular systems can assemble every system found in a cell.

Thus, nature demonstrates all the basic operations that are needed to perform molecular-level repairs on cells. What is more, as I described in Chapter 1, systems based on nanomachines will generally be more compact and capable than those found in nature. Natural systems show us only lower bounds to the possible, in cell repair as in everything else.

CELL REPAIR MACHINES

In short, with molecular technology and technical AI we will compile complete, molecular-level descriptions of healthy tissue, and we will build machines able to enter cells and to sense and modify their structures.

Cell repair machines will be comparable in size to bacteria and viruses, but their more-compact parts will allow them to be more complex. They will travel through tissue as white blood cells do, and enter cells as viruses do—or they could open and close cell membranes with a surgeon's care. Inside a cell, a repair machine will first size up the situation by examining the cell's contents and activity, and then take action. Early cell repair machines will be highly specialized, able to recognize and correct only a single type of molecular disorder, such as an enzyme deficiency or a form of DNA damage. Later machines (but not much later, with advanced technical AI systems doing the design work) will be programmed with more general abilities.

Complex repair machines will need nanocomputers to guide them. A micron-wide mechanical computer like that described in Chapter 1 will fit in 1/1000 of the volume of a typical cell, yet will hold more information than does the cell's DNA. In a repair system, such computers will direct smaller, simpler computers, which will in turn direct machines to examine, take apart, and rebuild damaged molecular structures.

By working along molecule by molecule and structure by structure, repair machines will be able to repair whole cells. By working along cell by cell and tissue by tissue, they (aided by larger devices, where need be) will be able to repair whole organs. By working through a person organ by organ, they will restore health. Because molecular machines will be able to build molecules and cells from scratch, they will be able to repair even cells damaged to the point of complete inactivity. Thus, cell repair machines will bring a fundamental breakthrough: they will free medicine from reliance on self-repair as the only path to healing.

To visualize an advanced cell repair machine, imagine it—and a cell—enlarged until atoms are the size of small marbles. On this scale, the repair machine's smallest tools have tips about the size of your fingertips; a medium-sized protein, like hemoglobin, is the size of a typewriter; and a ribosome is the size of a washing machine. A single repair device contains a simple computer the size of a small truck, along with many sensors of protein size, several manipulators of ribosome size, and provisions for memory and motive power. A total volume ten meters across, the size of a three-story house, holds all these parts and more. With parts the size of marbles packing this volume, the repair machine can do complex things.

But this repair device does not work alone. It, like its many siblings, is connected to a larger computer by means of mechanical data links the diameter of your arm. On this scale, a cubic-micron computer with a large memory fills a volume thirty stories high and as wide as a football field. The repair devices pass it information, and it passes back general instructions. Objects so large and complex are still small enough: on this scale, the cell itself is a kilometer across, holding one thousand times the volume of a cubic-micron computer, or a million times the volume of a single repair device. Cells are spacious.

Will such machines be able to do everything necessary to repair cells? Existing molecular machines demonstrate the ability to travel through tissue, enter cells, recognize molecular structures, and so forth, but other requirements are also important. Will repair machines work fast enough? If they do, will they waste so much power that the patient will roast?

The most extensive repairs cannot require *vastly* more work than building a cell from scratch. Yet molecular machinery working within a cellular volume routinely does just that, building a new cell in tens of minutes (in bacteria) to a few hours (in mammals). This indicates that repair machinery occupying a few percent of a cell's volume will be able to complete even extensive repairs in a reasonable time—days or weeks at most. Cells can spare this much room. Even brain cells can still function when an inert waste called lipofuscin (apparently a product of molecular damage) fills over ten percent of their volume.

Powering repair devices will be easy: cells naturally contain chemicals that power nanomachinery. Nature also shows that repair machines can be cooled: the cells in your body rework themselves steadily, and young animals grow swiftly without cooking themselves. Handling heat from a similar level of activity by repair machines will be no sweat—or at least not too much sweat, if a week of sweating is the price of health.

All these comparisons of repair machines to existing biological mechanisms raise the question of whether repair machines will be able to *improve* on nature. DNA repair provides a clear-cut illustration.

Just as an illiterate "book-repair machine" could recognize and repair a torn page, so a cell's repair enzymes can recognize and repair breaks and cross-links in DNA. Correcting misspellings (or mutations), though, would require an ability to read. Nature lacks such repair machines, but they will be easy to build. Imagine three identical DNA molecules, each with the same sequence of nucleotides. Now imagine each strand mutated to change a few scattered nucleotides. Each strand still seems normal, taken by itself. Nonetheless, a repair machine could compare each strand to the others, one segment at a time, and could note when a nucleotide failed to match its mates. Changing the odd nucleotide to match the other two will then repair the damage.

This method will fail if two strands mutate in the same spot. Imagine that the DNA of three human cells has been heavily damaged— after thousands of mutations, each cell has had one in every million nucleotides changed. The chance of our three-strand correction procedure failing at any given spot is then about one in a million million. But compare five strands at once, and the odds become about one in a million million million, and so on. A device that compares many strands will make the chance of an uncorrectable error effectively nil.

In practice, repair machines will compare DNA molecules from several cells, make corrected copies, and use these as standards for proofreading and repairing DNA throughout a tissue. By comparing several strands, repair machines will dramatically improve on nature's repair enzymes.

Other repairs will require different information about healthy cells and about how a particular damaged cell differs from the norm. Antibodies identify proteins by touch, and properly chosen antibodies can generally distinguish any two proteins by their differing shapes and surface properties. Repair machines will identify molecules in a similar way. With a suitable computer and data base, they will be able to identify proteins by reading their amino acid sequences.

Consider a complex and capable repair system. A volume of two cubic microns—about 2/1000 of the volume of a typical cell—will be enough to hold a central data base system able to:

1. Swiftly identify any of the hundred thousand or so different human proteins by examining a short amino acid sequence.

2. Identify all the other complex molecules normally found in cells.

3. Record the type and position of every large molecule in the cell.

Each of the smaller repair devices (of perhaps thousands in a cell) will include a less capable computer. Each of these computers will be able to perform over a thousand computational steps in the time that a typical enzyme takes to change a single molecular bond, so the speed of computation possible seems more than adequate. Because each computer will be in communication with a larger computer and the central data base, the available memory seems adequate. Cell repair machines will have both the molecular tools they need and "brains" enough to decide how to use them.

Such sophistication will be overkill (overcure?) for many health problems. Devices that merely recognize and destroy a specific kind of cell, for example, will be enough to cure a cancer. Placing a computer network in every cell may seem like slicing butter with a chain saw, but having a chain saw available does provide assurance that even hard butter can be sliced. It seems better to show too much than too little, if one aims to describe the limits of the possible in medicine.

SOME CURES

The simplest medical applications of nanomachines will involve not repair but selective destruction. Cancers provide one example; infectious diseases provide another. The goal is simple: one need only recognize and destroy the dangerous replicators, whether they are bacteria, cancer cells, viruses, or worms. Similarly, abnormal growths and deposits on arterial walls cause much heart disease; machines that recognize, break down, and dispose of them will clear arteries for more normal blood flow. Selective destruction will also cure diseases such as herpes in which a virus splices its genes into the DNA of a host cell. A repair device will enter the cell, read its DNA, and remove the addition that spells "herpes."

Repairing damaged, cross-linked molecules will also be fairly straightforward. Faced with a damaged, cross-linked protein, a cell repair machine will first identify it by examining short amino acid sequences, then look up its correct structure in a data base. The machine will then compare the protein to this blueprint, one amino acid at a time. Like a proofreader finding misspellings and strange characters (char#cters), it will find any changed amino acids or improper cross-links. By correcting these flaws, it will leave a normal protein, ready to do the work of the cell.

Repair machines will also aid healing. After a heart attack, scar tissue replaces dead muscle. Repair machines will stimulate the heart to grow fresh muscle by resetting cellular control mechanisms. By removing scar tissue and guiding fresh growth, they will direct the healing of the heart.

This list could continue through problem after problem (Heavy metal poisoning? —Find and remove the metal atoms) but the conclusion is easy to summarize. Physical disorders stem from misarranged atoms; repair machines will be able to return them to working order, restoring the body to health. Rather than compiling an endless list of curable diseases (from arthritis, bursitis, cancer, and dengue to yellow fever and zinc chills and back again), it makes sense to look for the limits to what cell repair machines can do. Limits do exist.

Consider stroke, as one example of a problem that damages the brain. Prevention will be straightforward: Is a blood vessel in the brain weakening, bulging, and apt to burst? Then pull it back into shape and guide the growth of reinforcing fibers. Does abnormal clotting threaten to block circulation? Then dissolve the clots and normalize the blood and blood-vessel linings to prevent a recurrence. Moderate neural damage from stroke will also be repairable: if reduced circulation has impaired function but left cell structures intact, then restore circulation and repair the cells, using their structures as a guide in restoring the tissue to its previous state. This will not only restore each cell's function, but will preserve the memories and skills embodied in the neural patterns in that part of the brain.

Repair machines will be able to regenerate fresh brain tissue even where damage has obliterated these patterns. But the patient would lose old memories and skills to the extent that they resided in that part of the brain. If unique neural patterns are truly obliterated, then cell repair machines could no more restore them than art conservators could restore a tapestry from stirred ash. Loss of information through obliteration of structure imposes the most important, fundamental limit to the repair of tissue.

Other tasks are beyond cell repair machines for different reasons—maintaining mental health, for instance. Cell repair machines will be able to correct some problems, of course. Deranged thinking sometimes has biochemical causes, as if the brain were drugging or poisoning itself, and other problems stem from tissue damage. But many problems have little to do with the health of nerve cells and everything to do with the health of the mind.

A mind and the tissue of its brain are like a novel and the paper of its book. Spilled ink or flood damage may harm the book, making the novel difficult to read. Book repair machines could nonetheless restore physical "health" by removing the foreign ink or by drying and repairing the damaged paper fibers. Such treatments would do nothing for the book's *content*, however, which in a real sense is nonphysical. If the book were a cheap romance with a moldy plot and empty characters, repairs would be needed not on the ink and paper, but on the novel. This would call not for physical repairs, but for more work by the author, perhaps with advice.

Similarly, removing poisons from the brain and repairing its nerve fibers will thin some mental fogs, but not revise the content of the mind. This can be changed by the patient, with effort; we are all authors of our minds. But because minds change themselves by changing their brains, having a healthy brain will aid sound thinking more than quality paper aids sound writing.

Readers familiar with computers may prefer to think in terms of hardware and software. A machine could repair a computer's hardware while neither understanding nor changing its software.

Such machines might stop the computer's activity but leave the patterns in memory intact and ready to work again. In computers with the right kind of memory (called "nonvolatile"), users do this by simply switching off the power. In the brain the job seems more complex, yet there could be medical advantages to inducing a similar state.

ANESTHESIA PLUS

Physicians already stop and restart consciousness by interfering with the chemical activity that underlies the mind. Throughout active life, molecular machines in the brain process molecules. Some disassemble sugars, combine them with oxygen, and capture the energy this releases. Some pump salt ions across cell membranes; others build small molecules and release them to signal other cells. Such processes make up the brain's metabolism, the sum total of its chemical activity. Together with its electrical effects, this metabolic activity underlies the changing patterns of thought.

Surgeons cut people with knives. In the mid-1800s, they learned to use chemicals that interfere with brain metabolism, blocking conscious thought and preventing patients from objecting so vigorously to being cut. These chemicals are anesthetics. Their molecules freely enter and leave the brain, allowing anesthetists to interrupt and restart human consciousness.

People have long dreamed of discovering a drug that interferes with the metabolism of the entire body, a drug able to interrupt metabolism completely for hours, days, or years. The result would be a condition of biostasis (from *bio,* meaning life, and *stasis,* meaning a stoppage or a stable state). A method of producing reversible biostasis could help astronauts on long space voyages to save food and avoid boredom, or it could serve as a kind of one-way time travel. In medicine, biostasis would provide a deep anesthesia giving physicians more time to work. When emergencies occur far from medical help, a good biostasis procedure would provide a sort of universal first-aid treatment: it would stabilize a patient's condition and prevent molecular machines from running amok and damaging tissues.

But no one has found a drug able to stop the entire metabolism the way anesthetics stop consciousness—that is, in a way that can be reversed by simply washing the drug out of the patient's tissues. Nonetheless, reversible biostasis will be possible when repair machines become available.

To see how one approach would work, imagine that the bloodstream carries simple molecular devices to tissues, where they enter the cells. There they block the molecular machinery of metabolism—in the brain and elsewhere—and tie structures together with stabilizing cross-links. Other molecular devices then move in, displacing water and packing themselves solidly around the molecules of the cell. These steps stop metabolism and preserve cell structures. Because cell repair machines will be used to reverse this process, it can cause moderate molecular damage and yet do no lasting harm. With metabolism stopped and cell structures held firmly in place, the patient will rest quietly, dreamless and unchanging, until repair machines restore active life.

If a patient in this condition were turned over to a present-day physician ignorant of the capabilities of cell repair machines, the consequences would likely be grim. Seeing no signs of life, the physician would likely conclude that the patient was dead, and then would make this judgment a reality by "prescribing" an autopsy, followed by burial or burning.

But our imaginary patient lives in an era when biostasis is known to be only an interruption of life, not an end to it. When the patient's contract says "wake me!" (or the repairs are complete, or the flight to the stars is finished), the attending physician begins resuscitation. Repair machines enter the patient's tissues, removing the packing from around the patient's molecules and replacing it with water. They then remove the cross-links, repair any damaged molecules and structures, and restore normal concentrations of salts, blood sugar, ATP, and so forth. Finally, they unblock the metabolic machinery. The interrupted metabolic processes resume, the patient yawns, stretches, sits up, thanks the doctor, checks the date, and walks out the door.

FROM FUNCTION TO STRUCTURE

The reversibility of biostasis and irreversibility of severe stroke damage help to show how cell repair machines will change medicine. Today, physicians can only help tissues to heal themselves. Accordingly, they must try to preserve the *function* of tissue. If tissues cannot function, they cannot heal. Worse, unless they are preserved, deterioration follows, ultimately obliterating structure. It is as if a mechanic's tools were able to work only on a running engine.

Cell repair machines change the central requirement from preserving *function* to preserving *structure*. As I noted in the discussion of stroke, repair machines will be able to restore brain function with memory and skills intact only if the distinctive structure of the neural fabric remains intact. Biostasis involves preserving neural structure while deliberately blocking function.

All this is a direct consequence of the molecular nature of the repairs. Physicians using scalpels and drugs can no more repair cells than someone using only a pickax and a can of oil can repair a fine watch. In contrast, having repair machines and ordinary nutrients will be like having a watchmaker's tools and an unlimited supply of spare parts. Cell repair machines will change medicine at its foundations.

FROM TREATING DISEASE
TO ESTABLISHING HEALTH

Medical researchers now study diseases, often seeking ways to prevent or reverse them by blocking a key step in the disease process. The resulting knowledge has helped physicians greatly: they now prescribe insulin to compensate for diabetes, anti-hypertensives to prevent stroke, penicillin to cure infections, and so on down an impressive list. Molecular machines will aid the study of diseases, yet they will make understanding disease far less important. Repair machines will make it more important to understand health.

The body can be ill in more ways than it can be healthy. Healthy muscle tissue, for example, varies in relatively few ways: it can be stronger or weaker, faster or slower, have this antigen or that one, and so forth. Damaged muscle tissue can vary in all these ways, yet also suffer from any combination of strains, tears, viral infections, parasitic worms, bruises, punctures, poisons, sarcomas, wasting diseases, and congenital abnormalities. Similarly, though neurons are woven in as many patterns as there are human brains, individual synapses and dendrites come in a modest range of forms—if they are healthy.

Once biologists have described normal molecules, cells, and tissues, properly programmed repair machines will be able to cure even unknown diseases. Once researchers describe the range of structures that (for example) a healthy liver may have, repair machines exploring a malfunctioning liver need only look for differences and correct them. Machines ignorant of a new poison and its effects will still recognize it as foreign and remove it. Instead of fighting a million strange diseases, advanced repair machines will establish a state of health.

Developing and programming cell repair machines will require great effort, knowledge, and skill. Repair machines with broad capabilities seem easier to build than to program. Their programs must contain detailed knowledge of the hundreds of kinds of cells and the hundreds of thousands of kinds of molecules in the human body. They must be able to map damaged cellular structures and decide how to correct them. How long will such machines and programs take to be developed? Offhand, the state of biochemistry and its present rate of advance might suggest that the basic knowledge alone will take centuries to collect. But we must beware of the illusion that advances will arrive in isolation.

Repair machines will sweep in with a wave of other technologies. The assemblers that build them will first be used to build instruments for analyzing cell structures. Even a pessimist might agree that human biologists and engineers equipped with these tools could build and program advanced cell repair machines in a hundred years of steady work. A cocksure, far-seeing pessimist might say a thousand years. A really committed nay-sayer might declare that the job would take people a million years. Very well: fast technical AI systems—a millionfold faster than scientists and engineers—will then develop advanced cell repair machines in a single calendar year.

A DISEASE CALLED "AGING"

Aging is natural, but so were smallpox and our efforts to prevent it. We have conquered smallpox, and it seems that we will conquer aging.

Longevity has increased during the last century, but chiefly because better sanitation and drugs have reduced bacterial illness. The basic human lifespan has increased little.

Still, researchers have made progress toward understanding and slowing the aging process. They have identified some of its causes, such as uncontrolled cross-linking. They have devised partial treatments, such as antioxidants and free-radical inhibitors. They have proposed and studied other mechanisms of aging, such as "clocks" in the cell and changes in the body's hormone balance. In laboratory experiments, special drugs and diets have extended the lifespan of mice by 25 to 45 percent.

Such work will continue; as the baby boom generation ages, expect a boom in aging research. One biotechnology company, Senetek of Denmark, specializes in aging research. In April 1985, Eastman Kodak and ICN Pharmaceuticals were reported to have joined in a $45 million venture to produce Isoprinosine and other drugs with the potential to extend lifespan. The results of conventional antiaging research may substantially lengthen human lifespans —and improve the health of the old—during the next ten to twenty years. How greatly will drugs, surgery, exercise, and diet extend lifespans? For now, estimates must remain guesswork. Only new scientific knowledge can rescue such predictions from the realm of speculation, because they rely on new science and not just new engineering.

With cell repair machines, however, the potential for life extension becomes clear. They will be able to repair cells so long as their distinctive structures remain intact, and will be able to replace cells that have been destroyed. Either way, they will restore health. Aging is fundamentally no different from any other physical disorder; it is no magical effect of calendar dates on a mysterious life-force. Brittle bones, wrinkled skin, low enzyme activities, slow wound healing, poor memory, and the rest all result from damaged molecular machinery, chemical imbalances, and misarranged structures. By restoring all the cells and tissues of the body to a youthful structure, repair machines will restore youthful health.

People who survive intact until the time of cell repair machines will have the opportunity to regain youthful health and to keep it almost as long as they please. Nothing can make a person (or anything else) last forever, of course, but barring severe accidents, those wishing to do so will live for a long, long time.

As a technology develops, there comes a time when its principles become clear, and with them many of its consequences. The principles of rocketry were clear in the 1930s, and with them the consequence of spaceflight. Filling in the details involved designing and testing tanks, engines, instruments, and so forth. By the early 1950s, many details were known. The ancient dream of flying to the Moon had became a goal one could plan for.

The principles of molecular machinery are already clear, and with them the consequence of cell repair machines. Filling in the details will involve designing molecular tools, assemblers, computers, and so forth, but many details of existing molecular machines are known today. The ancient dream of achieving health and long life has become a goal one can plan for.

Medical research is leading us, step by step, along a path toward molecular machinery. The global competition to make better materials, electronics, and biochemical tools is pushing us in the same direction. Cell repair machines will take years to develop, but they lie straight ahead.

They will bring many abilities, both for good and for ill. A moment's thought about military replicators with abilities like those of cell repair machines is enough to turn up nauseating possibilities. Later I will describe how we might avoid such horrors, but it first seems wise to consider the alleged benefits of cell repair machines. Is their *apparent* good *really* good? How might long life affect the world?

8

Long Life in an Open World

The long habit of living indisposeth us for dying.

—Sir THOMAS BROWNE

CELL REPAIR MACHINES raise questions involving the value of extending human life. These are not the questions of today's medical ethics, which commonly involve dilemmas posed by scarce, costly, and half-effective treatments. They are instead questions involving the value of long, healthy lives achieved by inexpensive means.

For people who value human life and enjoy living, such questions may need no answer. But after a decade marked by concern about population growth, pollution, and resource depletion, many people may question the desirability of extending life; such concerns have fostered the spread of pro-death memes. These memes must be examined afresh, because many have roots in an obsolete worldview. Nanotechnology will change far more than just human lifespan.

We will gain the means not only to heal ourselves, but to heal Earth of the wounds we have inflicted. Since saving lives will increase the number of the living, life extension raises questions about the effect of more people. Our ability to heal the Earth will lessen one cause for controversy.

Still, cell repair machines themselves will surely stir controversy. They disturb traditional assumptions about our bodies and our futures: this makes doubt soothing. They will require several major breakthroughs: this makes doubt easy. Since the possibility or impossibility of cell repair machines raises important issues, it makes sense to consider what objections might be raised.

WHY *NOT* CELL REPAIR MACHINES?

What sort of argument could suggest that cell repair machines are impossible? A successful argument must manage some strange contortions. It must somehow hold that molecular machines cannot build and repair cells, while granting that the molecular machines in our bodies actually *do* build and repair cells every day. A cruel problem for the committed skeptic! True, artificial machines must do what natural machines fail to do, but they need not do anything *qualitatively* novel. Both natural and artificial repair devices must reach, identify, and rebuild molecular structures. We will be able to improve on existing DNA repair enzymes simply by comparing several DNA strands at once, so nature obviously hasn't found all the tricks. Since this example explodes any general argument that repair machines cannot improve on nature, a good case against cell repair machines seems difficult to make.

Still, two general questions deserve direct answers. First, why should we expect to achieve long life in the coming decades, when people have tried and failed for millennia? Second, if we can indeed use cell repair machines to extend lives, then why hasn't nature (which has been repairing cells for billions of years) already perfected them?

People have tried and failed.

For centuries, people have longed to escape their short lifespans. Every so often, a Ponce de León or a quack doctor has promised a potion, but it has never worked. These statistics of failure have persuaded some people that, since all attempts have failed, all always will fail. They say "Aging is natural," and to them that seems reason enough. Medical advances may have shaken their views, but advances have chiefly reduced early death, not extended maximum life-span.

But now biochemists have gone to work examining the machines that build, repair, and control cells. They have learned to assemble viruses and reprogram bacteria. For the first time in history, people are examining their molecules and unraveling the molecular secrets of life. It seems that molecular engineers will eventually combine improved biochemical knowledge with improved molecular machines, learning to repair damaged tissue structures and so rejuvenate them. This is nothing strange—it would be strange, rather, if such powerful knowledge and abilities did *not* bring dramatic results. The massive statistics of past failure are simply irrelevant, because we have never before tried to build cell repair machines.

Nature has tried and failed.

Nature *has* been building cell repair machines. Evolution has tinkered with multicelled animals for hundreds of millions of years, yet advanced animals all age and die, because nature's nanomachines repair cells imperfectly. Why should improvements be possible?

Rats mature in months, and then age and die in about two years—yet human beings have evolved to live over thirty times longer. If longer lives were the chief goal of evolution, then rats would live longer too. But durability has costs: to repair cells requires an investment in energy, materials, and repair machines. Rat genes direct rat bodies to invest in swift growth and reproduction, not in meticulous self-repair. A rat that dallied in reaching breeding size would run a greater risk of becoming a cat snack first. Rat genes have prospered by treating rat bodies as cheap throwaways. Human genes likewise discard human beings, though after a life a few dozen times longer than a rat's.

But shoddy repairs are not the only cause of aging. Genes turn egg cells into adults through a pattern of development which rolls forward at fairly steady speed. This pattern is fairly consistent because evolution seldom changes a basic design. Just as the basic pattern of the DNA-RNA-protein system froze several billions of years ago, so the basic pattern of chemical signals and tissue responses that guides mammalian development jelled many millions of years ago. That process apparently has a clock, set to run at different speeds in different species, and a program that runs out.

Whatever the causes of aging, evolution has had little reason to eliminate them. If genes built individuals able to stay healthy for millennia, they would gain little advantage in their "effort" to replicate. Most individuals would still die young from starvation, predation, accident, or disease. As Sir Peter Medawar points out, a gene that helps the young (who are many) but harms the old (who are few) will replicate well and so spread through the population. If enough such genes accumulate, animals become programmed to die.

Experiments by Dr. Leonard Hayflick suggest that cells contain "clocks" that count cell divisions and stop the division process when the count gets too high. A mechanism of this sort can help young animals: if cancer-like changes make a cell divide too rapidly, but fail to destroy its clock, then it will grow to a tumor of limited size. The clock would thus prevent the unlimited growth of a true cancer. Such clocks could harm older animals by stopping the division of normal cells, ending tissue renewal. The animal thus would benefit from reduced cancer rates when young, yet have cause to complain if it lives to grow old. But its genes won't listen—they will have jumped ship earlier, as copies passed to the next generation. With cell repair machines we will be able to reset such clocks. Nothing suggests that evolution has perfected our bodies even by the brute standard of survival and reproduction. Engineers don't wire computers with slow, nervelike fibers or build machines out of soft protein, and for good reason. Genetic evolution (unlike memetic evolution) has been unable to leap to new materials or new systems, but has instead refined and extended the old ones.

The cell's repair machines fall far short of the limits of the possible —they don't even have computers to direct them. The lack of nano-computers in cells, of course, shows only that computers couldn't (or simply *didn't*) evolve gradually from other molecular machines. Nature has failed to build the best possible cell repair machines, but there have been ample reasons.

HEALING AND PROTECTING THE EARTH

The failure of Earth's biological systems to adapt to the industrial revolution is also easy to understand. From deforestation to dioxin, we have caused damage faster than evolution can respond. As we have sought more food, goods, and services, our use of bulk technology has forced us to continue such damage. With future technology, though, we will be able to do more good for ourselves, yet do less harm to the Earth. In addition, we will be able to build planet-mending machines to correct damage already done. Cells are not all we will want to repair.

Consider the toxic waste problem. Whether in our air, soil, or water, wastes concern us because they can harm living systems. But any materials that come in contact with the molecular machinery of life can themselves be reached by other forms of molecular machinery. This means that we will be able to design cleaning machines to remove these poisons wherever they could harm life.

Some wastes, such as dioxin, consist of dangerous molecules made of innocuous atoms. Cleaning machines will render them harmless by rearranging their atoms. Other wastes, such as lead and radioactive isotopes, contain dangerous atoms. Cleaning machines will collect these for disposal in any one of several ways. Lead comes from Earth's rocks; assemblers could build it into rocks in the mines from which it came. Radioactive isotopes could also be isolated from living things, either by building them into stable rock or by more drastic means. Using cheap, reliable space transportation systems, we could bury them in the dead, dry rock of the Moon. Using nanomachines, we could seal them in self-repairing, self-sealing containers the size of hills and powered by desert sunlight. These would be more secure than any passive rock or cask.

With replicating assemblers, we will even be able to remove the billions of tons of carbon dioxide that our fuel-burning civilization has dumped into the atmosphere. Climatologists project that climbing carbon dioxide levels, by trapping solar energy, will partially melt the polar caps, raising sea levels and flooding coasts sometime in the middle of the next century. Replicating assemblers, though, will make solar power cheap enough to eliminate the need for fossil fuels. Like trees, solar-powered nanomachines will be able to extract carbon dioxide from the air and split off the oxygen. Unlike trees, they will be able to grow deep storage roots and place carbon back in the coal seams and oil fields from which it came.

Future planet-healing machines will also help us mend torn landscapes and restore damaged ecosystems. Mining has scraped and pitted the Earth; carelessness has littered it. Fighting forest fires has let undergrowth thrive, replacing the cathedral-like openness of ancient forests with scrub growth that feeds more dangerous fires. We will use inexpensive, sophisticated robots to reverse these effects and others. Able to move rock and soil, they will recontour torn lands. Able to weed and digest, they will simulate the clearing effects of natural forest fires without danger or devastation. Able to lift and move trees, they will thin thick stands and reforest bare hills. We will make squirrel-sized devices with a taste for old trash. We will make treelike devices with roots that spread deep and cleanse the soil of pesticides and excess acid. We will make insect-sized lichen cleaners and spray-paint nibblers. We will make whatever devices we need to clean up the mess left by twentieth-century civilization.

After the cleanup, we will recycle most of these machines, keeping only those we still need to protect the environment from a cleaner civilization based on molecular technology. These more lasting devices will supplement natural ecosystems wherever needed, to balance and heal the effects of humanity. To make them effective, harmless, and hidden will be a craft requiring not just automated engineering, but knowledge of nature and a sense of art.

With cell repair technology, we will even be able to return some species from apparent extinction. The African quagga—a zebralike animal—became extinct over a century ago, but a salt-preserved quagga pelt survived in a German museum. Alan Wilson of the University of California at Berkeley and his co-workers have used enzymes to extract DNA fragments from muscle tissue attached to this pelt. They cloned the fragments in bacteria, compared them to zebra DNA, and found (as expected) that the genes showed a close evolutionary relationship. They have also succeeded in extracting and replicating DNA from a century-old bison pelt and from millennia-old mammoths preserved in the arctic permafrost. This success is a far cry from cloning a whole cell or organism—cloning one gene leaves about 100,000 uncloned, and cloning every gene still doesn't repair a single cell—but it does show that the hereditary material of these species still survives.

As I described in the last chapter, machines that compare several damaged copies of a DNA molecule will be able to reconstruct an undamaged original—and the billions of cells in a dried skin contain *billions* of copies. From these, we will be able to reconstruct undamaged DNA, and around the DNA we will be able to construct undamaged cells of whatever type we desire. Some insect species pass through winter as egg cells, to be revived by the warmth of spring. These "extinct" species will pass through the twentieth century as skin and muscle cells, to be converted into fertile eggs and revived by cell repair machines.

Dr. Barbara Durrant, a reproductive physiologist at the San Diego Zoo, is preserving tissue samples from endangered species in a cryogenic freezer. The payoff may be greater than most people now expect. Preserving just tissue samples doesn't preserve the life of an animal or an ecosystem, but it does preserve the genetic heritage of the sampled species. We would be reckless if we failed to take out this insurance policy against the permanent loss of species. The prospect of cell repair machines thus affects our choices today.

Extinction is not a new problem. About 65 million years ago, most then-existing species vanished, including all species of dinosaur. In Earth's book of stone, the story of the dinosaurs ends on a page consisting of a thin layer of clay. The clay is rich in iridium, an element common in asteroids and comets. The best current theory indicates that a blast from the sky smashed Earth's biosphere. With the energy of a hundred million megatons of TNT, it spread dust and an "asteroidal winter" planetwide.

In the eons since living cells first banded together to form worms, Earth has suffered five great extinctions. Only 34 million years ago—some 30 million years after the dinosaurs died—a layer of glassy beads settled to the seafloor. Above that layer the fossils of many species vanish. These beads froze from the molten splash of an impact.

Meteor Crater, in Arizona, bears witness to a smaller, more recent blast equaling that of a four-megaton bomb. As recently as June 30, 1908, a ball of fire split the Siberian sky and blasted the forest flat across an area a hundred kilometers wide.

As people have long suspected, the dinosaurs died because they were stupid. Not that they were too stupid to feed, walk, or guard their eggs—they did survive for 140 million years—they were merely too stupid to build telescopes able to detect asteroids and spacecraft able to deflect them from collision with Earth. Space has more rocks to throw at us, but we are showing signs of adequate intelligence to deal with them. When nanotechnology and automated engineering give us a more capable space technology, we will find it easy to track and deflect asteroids; in fact, we could do it with technology available today. We can both heal Earth and protect it.

LONG LIFE AND POPULATION PRESSURE

People commonly seek long, healthy lives, yet the prospect of a dramatic success is unsettling. Might greater longevity harm the quality of life? How will the prospect of long life affect our immediate problems? Though most effects cannot be foreseen, some can.

For example, as cell repair machines extend life, they will increase population. If all else were equal, more people would mean greater crowding, pollution, and scarcity—but all else will *not* be equal: the very advances in automated engineering and nanotechnology that will bring cell repair machines will also help us heal the Earth, protect it, and live more lightly upon it. We will be able to produce our necessities and luxuries without polluting our air, land, or water. We will be able to get resources and make things without scarring the landscape with mines or cluttering it with factories. With efficient assemblers making durable products, we will produce things of greater value with less waste. More people will be able to live on Earth, yet do less harm to it—or to one another, if we somehow manage to use our new abilities for good ends.

If one were to see the night sky as a black wall and expect the technology race to screech to a polite halt, then it would be natural to fear that long-lived people would be a burden on the "poor, crowded world of our children." This fear stems from the illusion that life is a zero-sum game, that having more people always means slicing a small pie thinner. But when we become able to repair cells, we will also be able to build replicating assemblers and excellent spacecraft. Our "poor" descendants will share a world the size of the solar system, with matter, energy, and potential living space dwarfing our entire planet.

This will open room enough for an era of growth and prosperity far beyond any precedent. Yet the solar system itself is finite, and the stars are distant. On Earth, even the cleanest assembler-based industries will produce waste heat. Concern about population and resources will remain important because the exponential growth of replicators (such as people) can eventually overrun any finite resource base.

But does this mean that we should sacrifice lives to delay the crunch? A few people may volunteer themselves, but they will do little good. In truth, life extension will have little effect on the basic problem: exponential growth will remain exponential whether people die young or live indefinitely. A martyr, by dying early, could delay the crisis by a fraction of a second—but a halfway dedicated person could help more by joining a movement of long-lived people working to solve this long-range problem. After all, many people have ignored the limits to growth on Earth. Who but the long-lived will prepare for the firmer but more distant limits to growth in the world beyond Earth? Those concerned with long-term limits will serve humanity best by staying alive, to keep their concern alive.

Long life also raises the threat of cultural stagnation. If this were an inevitable problem of long life, it is unclear what one could do about it—machine-gun the old for holding firm opinions, perhaps? Fortunately, two factors will reduce the problem somewhat. First, in a world with an open frontier the young will be able to move out, build new worlds, test new ideas, and then either persuade their elders to change or leave them behind. Second, people old in years will be young in body and brain. Aging slows both learning and thought, as it slows other physical processes; rejuvenation will speed them again. Since youthful muscles and sinews make young bodies more flexible, perhaps youthful brain tissues will keep minds somewhat more flexible, even when steeped in long years of wisdom.

THE EFFECTS OF ANTICIPATION

Long life will not be the greatest of the future's problems. It might even help solve them.

Consider its effect on people's willingness to start wars. Aging and death have made slaughter in combat more acceptable: As Homer had Sarpedon, hero of Troy, say, "O my friend, if we, leaving this war, could escape from age and death, I should not here be fighting in the van; but now, since many are the modes of death impending over us which no man can hope to shun, let us press on and give renown to other men, or win it for ourselves."

Yet if the hope of escaping age and death turns people from battle, will this be good? It might discourage small wars that could grow into a nuclear holocaust. But equally, it might weaken our resolve to defend ourselves from lifelong oppression—if we take no account of how much more life we have to defend. The reluctance of others to die for their ruler's power will help.

Expectations always shape actions. Our institutions and personal plans both reflect our expectation that all adults now living will die in mere decades. Consider how this belief inflames the urge to acquire, to ignore the future in pursuit of a fleeting pleasure. Consider how it blinds us to the future, and obscures the long-term benefits of cooperation. Erich Fromm writes: "If the individual lived five hundred or one thousand years, this clash (between his interests and those of society) might not exist or at least might be considerably reduced. He then might live and harvest with joy what he sowed in sorrow; the suffering of one historical period which will bear fruit in the next one could bear fruit for him too." Whether or not most people will still live for the present is beside the point: the question is, might there be a *significant* change for the better?

The expectation of living a long life in a better future may well make some political diseases less deadly. Human conflicts are far too deep and strong to be uprooted by any simple change, yet the prospect of vast wealth tomorrow may at least *lessen* the urge to fight over crumbs today. The problem of conflict is great, and we need all the help we can get.

The prospect of personal deterioration and death has always made thoughts of the future less pleasant. Visions of pollution, poverty, and nuclear annihilation have recently made thoughts of the future almost too gruesome to bear. Yet with at least a hope of a better future and time to enjoy it, we may look forward more willingly. Looking forward, we will see more. Having a personal stake, we will care more. Greater hope and foresight will benefit both the present and posterity; they will even better our odds of survival.

Lengthened lives will mean more people, but without greatly worsening tomorrow's population problem. The expectation of longer lives in a better world will bring real benefits, by encouraging people to give more thought to the future. Overall, long life and its anticipation seem good for society, just as shortening lifespans to thirty would be bad. Many people want long, healthy lives for themselves. What are the prospects for the present generation?

PROGRESS IN LIFE EXTENSION

Hear Gilgamesh, King of Uruk:

"I have looked over the wall and I see the bodies floating on the river, and that will be my lot also. Indeed I know it is so, for whoever is tallest among men cannot reach the heavens, and the greatest cannot encompass the earth."

Four millennia have passed since Sumerian scribes marked clay tablets to record *The Epic of Gilgamesh,* and times have changed. Men no taller than average have now reached the heavens and circled the Earth. We of the Space Age, the Biotechnology Age, the Age of Breakthroughs—need we still despair before the barrier of years? Or will we learn the art of life extension soon enough to save ourselves and those we love from dissolution?

The pace of biomedical advance holds tantalizing promise. The major diseases of age—heart disease, stroke, and cancer—have begun to yield to treatment. Studies of aging mechanisms have begun to bear fruit, and researchers have extended animals' lifespans. As knowledge builds on knowledge and tools lead to new tools, advances seem sure to accelerate. Even without cell repair machines, we have reason to expect major progress toward slowing and partially reversing aging.

Although people of all ages will benefit from these advances, the young will benefit more. Those surviving long enough will reach a time when aging becomes fully reversible: at the latest, the time of advanced cell repair machines. Then, if not sooner, people will grow healthier as they grow older, improving like wine instead of spoiling like milk. They will, if they choose, regain excellent health and live a long, long time.

In that time, with its replicators and cheap spaceflight, people will have both long lives and room and resources enough to enjoy them. A question that may roll bitterly off the tongue is: "When? . . . Which will be the last generation to age and die, and which the first to win through?" Many people now share the quiet expectation that aging will someday be conquered. But are those now alive doomed by a fluke of premature birth? The answer will prove both clear and startling.

The obvious path to long life involves living long enough to be rejuvenated by cell repair machines. Advances in biochemistry and molecular technology will extend life, and in the time won they will extend it yet more. At first we will use drugs, diet, and exercise to extend healthy life. Within several decades, advances in nanotechnology will likely bring early cell repair machines—and with the aid of automated engineering, early machines may promptly be followed by advanced machines. Dates must remain mere guesses, but a guess will serve better than a simple question mark.

Imagine someone who is now thirty years old. In another thirty years, biotechnology will have advanced greatly, yet that thirty-year-old will be only sixty. Statistical tables which assume *no* advances in medicine say that a thirty-year-old U.S. citizen can now expect to live almost *fifty* more years—that is, well into the 2030s. Fairly routine advances (of sorts demonstrated in animals) seem likely to add years, perhaps decades, to life by 2030. The mere beginnings of cell repair technology might extend life by several decades. In short, the medicine of 2010, 2020, and 2030 seems likely to extend our thirty-year-old's life into the 2040s or 2050s. By then, if not before, medical advances may permit actual rejuvenation. Thus, those under thirty (and perhaps those substantially older) can look forward—at least tentatively—to medicine's overtaking their aging process and delivering them safely to an era of cell repair, vigor, and indefinite lifespan.

If this were the whole story, then the division between the last on the road to early death and the first on the road to long life would be perhaps the ultimate gap between generations. What is more, a gnawing uncertainty about one's own fate would give reason to push the whole matter into the subconscious dungeon of disturbing speculations.

But is this really our situation? There seems to be another way to save lives, one based on cell repair machines, yet applicable today. As the last chapter described, repair machines will be able to heal tissue so long as its essential structure is preserved. A tissue's ability to metabolize and to repair itself becomes unimportant; the discussion of biostasis illustrated this. Biostasis, as described, will use molecular devices to stop function and preserve structure by cross-linking the cell's molecular machines to one another. Nanomachines will reverse biostasis by repairing molecular damage, removing cross-links, and helping cells (and hence tissues, organs, and the whole body) return to normal function.

Reaching an era with advanced cell repair machines seems the key to long life and health, because almost all physical problems will then be curable. One might manage to arrive in that era by remaining alive and active through all the years between now and then—but this is merely the most obvious way, the way that requires a minimum of foresight. Patients today often suffer a collapse of heart function while the brain structures that embody memory and personality remain intact. In such cases, might not *today's* medical technology be able to stop biological processes in a way that *tomorrow's* medical technology will be able to reverse? If so, then most deaths are now prematurely diagnosed, and needless.

9

A Door to the Future

London, April 1773.

To Jacques Dubourg.

Your observations on the causes of death, and the experiments which you propose for recalling to life those who appear to be killed by lightning, demonstrate equally your sagacity and your humanity. It appears that the doctrine of life and death in general is yet but little understood . . .

I wish it were possible . . . to invent a method of embalming drowned persons, in such a manner that they might be recalled to life at any period, however distant; for having a very ardent desire to see and observe the state of America a hundred years hence, I should prefer to an ordinary death, being immersed with a few friends in a cask of Madeira, until that time, then to be recalled to life by the solar warmth of my dear country! But . . . in all probability, we live in a century too little advanced, and too near the infancy of science, to see such an art brought in our time to its perfection . . .

I am, etc.
B. FRANKLIN.

BENJAMIN FRANKLIN wanted a procedure for stopping and re-
starting metabolism, but none was then known. Do we live in a
century far enough advanced to make biostasis available—to open a
future of health to patients who would otherwise lack any choice but
dissolution after they have expired?

We can stop metabolism in many ways, but biostasis, to be of use,
must be reversible. This leads to a curious situation. Whether we can
place patients in biostasis using *present* techniques depends entirely
on whether *future* techniques will be able to reverse the process. The
procedure has two parts, of which we must master only one.

If biostasis can keep a patient unchanged for years, then those
future techniques will include sophisticated cell repair systems. We
must therefore judge the success of present biostasis procedures in
light of the ultimate abilities of future medicine. Before cell repair
machines became a clear prospect, those abilities—and thus the requirements for successful biostasis—remained grossly uncertain.
Now, the basic requirements seem fairly obvious.

THE REQUIREMENTS FOR BIOSTASIS

Molecular machines can build cells from scratch, as dividing cells
demonstrate. They can also build organs and organ systems from
scratch, as developing embryos demonstrate. Physicians will be able
to use cell repair technology to direct the growth of new organs from
a patient's own cells. This gives modern physicians great leeway in
biostasis procedures: even if they were to damage or discard most of
a patient's organs, they would still do no irreversible harm. Future
colleagues with better tools will be able to repair or replace the
organs involved. Most people would be glad to have a new heart,
fresh kidneys, or younger skin.

But the brain is another matter. A physician who allows the destruction of a patient's brain allows the destruction of the patient as a
person, whatever may happen to the rest of the body. The brain
holds the patterns of memory, of personality, of self. Stroke patients
lose only parts of their brains, yet suffer harm ranging from partial
blindness to paralysis to loss of language, lowered intelligence, altered personality, and worse. The effects depend on the location of
the damage. This suggests that total destruction of the brain causes
total blindness, paralysis, speechlessness, and mindlessness, whether
the body continues to breathe or not.

As Voltaire wrote, "To rise again—to be the same person that you were—you must have your memory perfectly fresh and present; for it is memory that makes your identity. If your memory be lost, how will you be the same man?" Anesthesia interrupts consciousness without disrupting the structure of the brain, and biostasis procedures must do likewise, for a longer time. This raises the question of the nature of the physical structures that underlie memory and personality.

Neurobiology, and informed common sense, agree on the basic nature of memory. As we form memories and develop as individuals, our brains change. These changes affect the brain's function, changing its pattern of activity: When we remember, our brains *do* something; when we act, think, or feel, our brains *do* something. Brains work by means of molecular machinery. Lasting changes in brain function involve lasting changes in this molecular machinery—unlike a computer's memory, the brain is not designed to be wiped clean and refilled at a moment's notice. Personality and long-term memory are durable.

Throughout the body, durable changes in function involve durable changes in molecular machinery. When muscles become stronger or swifter, their proteins change in number and distribution. When a liver adapts to cope with alcohol, its protein content also changes. When the immune system learns to recognize a new kind of influenza virus, protein content changes again. Since protein-based machines do the actual work of moving muscles, breaking down toxins, and recognizing viruses, this relationship is to be expected.

In the brain, proteins shape nerve cells, stud their surfaces, link one cell to the next, control the ionic currents of each neural impulse, produce the signal molecules that nerve cells use to communicate across synapses, and much, much more. When printers print words, they put down patterns of ink; when nerve cells change their behavior, they change their patterns of protein. Printing also dents the paper, and nerve cells change more than just their proteins, yet the ink on the paper and the proteins in the brain are enough to make these patterns clear. The changes involved are far from subtle. Researchers report that long-term changes in nerve cell behavior involve "striking morphological changes" in synapses: they change visibly in size and structure.

It seems that long-term memory is not some terribly delicate pattern, ready to evaporate from the brain at any excuse. Memory and personality are instead firmly embodied in the way that brain cells have grown together, in patterns formed through years of experience. Memory and personality are no more material than the characters in a novel; yet like them they are embodied in matter. Memory and personality do not waft away on the last breath as a patient expires. Indeed, many patients have recovered from so-called "clinical death," even without cell repair machines to help. The patterns of mind are destroyed only when and if the attending physicians allow the patient's brain to undergo dissolution. This again allows physicians considerable leeway in biostasis procedures: typically, they need not stop metabolism until after vital functions have ceased.

It seems that preserving the cell structures and protein patterns of the brain will also preserve the structure of the mind and self. Biologists already know how to preserve tissue this well. Resuscitation technology must await cell repair machines, but biostasis technology seems well in hand.

METHODS OF BIOSTASIS

The idea that we already have biostasis techniques may seem surprising, since powerful new abilities seldom spring up overnight. In fact, the techniques are old—only understanding of their reversibility is new. Biologists developed the two main approaches for other reasons.

For decades, biologists have used electron microscopes to study the structure of cells and tissues. To prepare specimens, they use a chemical process called *fixation* to hold molecular structures in place. A popular method uses glutaraldehyde molecules, flexible chains of five carbon atoms with a reactive group of hydrogen and oxygen atoms at each end. Biologists fix tissue by pumping a glutaraldehyde solution through blood vessels, which allows glutaraldehyde molecules to diffuse into cells. A molecule tumbles around inside a cell until one end contacts a protein (or other reactive molecule) and bonds to it. The other end then waves free until it, too, contacts something reactive. This commonly shackles a protein molecule to a neighboring molecule.

These cross-links lock molecular structures and machines in place; other chemicals then can be added to do a more thorough or sturdy job. Electron microscopy shows that such fixation procedures preserve cells and the structures within them, including the cells and structures of the brain.

The first step of the hypothetical biostasis procedure that I described in Chapter 7 involved simple molecular devices able to enter cells, block their molecular machinery, and tie structures together with stabilizing cross-links. Glutaraldehyde molecules fit this description quite well. The next step in this procedure involved other molecular devices able to displace water and pack themselves solidly around the molecules of a cell. This also corresponds to a known process.

Chemicals such as propylene glycol, ethylene glycol, and dimethyl sulfoxide can diffuse into cells, replacing much of their water yet doing little harm. They are known as "cryoprotectants," because they can protect cells from damage at low temperatures. If they replace enough of a cell's water, then cooling doesn't cause freezing, it just causes the protectant solution to become more and more viscous, going from a liquid that resembles thin syrup in its consistency to one that resembles hot tar, to one that resembles cold tar, to one as resistant to flow as a glass. In fact, according to the scientific definition of the term, the protectant solution then qualifies as a glass; the process of solidification without freezing is called vitrification. Mouse embryos vitrified and stored in liquid nitrogen have grown into healthy mice.

The vitrification process packs the glassy protectant solidly around the molecules of each cell; vitrification thus fits the description I gave of the second stage of biostasis.

Fixation and vitrification together seem adequate to ensure long-term biostasis. To reverse this form of biostasis, cell repair machines will be programmed to remove the glassy protectant and the glutaraldehyde cross-links and then repair and replace molecules, thus restoring cells, tissues, and organs to working order.

Fixation with vitrification is not the first procedure proposed for biostasis. In 1962 Robert Ettinger, a professor of physics at Highland Park College in Michigan, published a book suggesting that future advances in cryobiology might lead to techniques for the easily reversible freezing of human patients. He further suggested that physicians using future technology might be able to repair and revive patients frozen with present techniques shortly after cessation of vital signs. He pointed out that liquid nitrogen temperatures will preserve patients for centuries, if need be, with little change. Perhaps, he suggested, medical science will one day have "fabulous machines" able to restore frozen tissue a molecule at a time. His book gave rise to the cryonics movement.

Cryonicists have focused on freezing because many human cells revive *spontaneously* after careful freezing and thawing. It is a common myth that freezing bursts cells; in fact, freezing damage is more subtle than this—so subtle that it often does no lasting harm. Frozen sperm regularly produces healthy babies. Some human beings now alive have survived being frozen solid at liquid nitrogen temperatures—when they were early embryos. Cryobiologists are actively researching ways to freeze and thaw viable organs to allow surgeons to store them for later implantation.

The prospect of future cell repair technologies has been a consistent theme among cryonicists. Still, they have tended to focus on procedures that preserve cell function, for natural reasons. Cryobiologists have kept viable human cells frozen for years. Researchers have improved their results by experimenting with mixes of cryoprotective chemicals and carefully controlled cooling and warming rates. The complexities of cryobiology offer rich possibilities for further experimentation. This combination of tangible, tantalizing success and promising targets for further research has made the quest for an easily reversible freezing process a vivid and attractive goal for cryonicists. A success at freezing and reviving an adult mammal would be immediately visible and persuasive.

What is more, even *partial* preservation of tissue function suggests *excellent* preservation of tissue structure. Cells that can revive (or almost revive) even without special help will need little repair.

The cryonics community's cautious, conservative emphasis on preserving tissue function has invited public confusion, though. Experimenters have frozen whole adult mammals and thawed them without waiting for the aid of cell repair machines. The results have been superficially discouraging: the animals fail to revive. To a public and a medical community that has known nothing about the prospects for cell repair, this has made frozen biostasis seem pointless.

And, after Ettinger's proposal, a few cryobiologists chose to make unsupported pronouncements about the future of medical technology. As Robert Prehoda stated in a 1967 book: "Almost all reduced-metabolism experts . . . believe that cellular damage caused by current freezing techniques could never be corrected." Of course, these were the wrong experts to ask. The question called for experts on molecular technology and cell repair machines. These cryobiologists should have said only that correcting freezing damage would apparently require molecular-level repairs, and that they, personally, had not studied the matter. Instead, they casually misled the public on a matter of vital medical importance. Their statements discouraged the use of a workable biostasis technique.

Cells are mostly water. At low enough temperatures, water molecules join to form a weak but solid framework of cross-links. Since this preserves neural structures and thus the patterns of mind and memory, Robert Ettinger has apparently identified a workable approach to biostasis. As molecular technology advances and people grow familiar with its consequences, the reversibility of biostasis (whether based on freezing, fixation and vitrification, or other methods) will grow ever more obvious to ever more people.

REVERSING BIOSTASIS

Imagine that a patient has expired because of a heart attack. Physicians attempt resuscitation but fail, and give up on restoring vital functions. At this point, though, the patient's body and brain are just *barely* nonfunctional—most cells and tissues, in fact, are still alive and metabolizing. Having made arrangements beforehand, the patient is soon placed in biostasis to prevent irreversible dissolution and await a better day.

Years pass. The patient changes little, but technology advances greatly. Biochemists learn to design proteins. Engineers use protein machines to build assemblers, then use assemblers to build a broad-based nanotechnology. With new instruments, biological knowledge explodes. Biomedical engineers use new knowledge, automated engineering, and assemblers to develop cell repair machines of growing sophistication. They learn to stop and reverse aging. Physicians use cell repair technology to resuscitate patients in biostasis—first those placed in biostasis by the most advanced techniques, then those placed in biostasis using earlier and cruder techniques. Finally, after the successful resuscitation of animals placed in biostasis using the old techniques of the 1980s, physicians turn to our heart-attack patient.

In the first stage of preparation, the patient lies in a tank of liquid nitrogen surrounded by equipment. Glassy protectant still locks each cell's molecular machinery in a firm embrace. This protectant must be removed, but simple warming might allow some cell structures to move about prematurely.

Surgical devices designed for use at low temperatures reach through the liquid nitrogen to the patient's chest. There they remove solid plugs of tissue to open access to major arteries and veins. An army of nanomachines equipped for removing protectant moves through these openings, clearing first the major blood vessels and then the capillaries. This opens paths throughout the normally active tissues of the patient's body. The larger surgical machines then attach tubes to the chest and pump fluid through the circulatory system. The fluid washes out the initial protectant-removal machines (later, it supplies materials to repair machines and carries away waste heat).

Now the machines pump in a milky fluid containing trillions of devices that enter cells and remove the glassy protectant, molecule by molecule. They replace it with a temporary molecular scaffolding that leaves ample room for repair machines to work. As these protectant-removal machines uncover biomolecules, including the structural and mechanical components of the cells, they bind them to the scaffolding with temporary cross-links. (If the patient had also been treated with a cross-linking fixative, these cross-links would now be removed and replaced with the temporary links.) When molecules must be moved aside, the machines label them for proper replacement. Like other advanced cell repair machines, these devices work under the direction of on-site nanocomputers.

When they finish, the low-temperature machines withdraw. Through a series of gradual changes in composition and temperature, a water-based solution replaces the earlier cryogenic fluid and the patient warms to above the freezing point. Cell repair machines are pumped through the blood vessels and enter the cells. Repairs commence.

Small devices examine molecules and report their structures and positions to a larger computer within the cell. The computer identifies the molecules, directs any needed molecular repairs, and identifies cell structures from molecular patterns. Where damage has displaced structures in a cell, the computer directs the repair devices to restore the molecules to their proper arrangement, using temporary cross-links as needed. Meanwhile, the patient's arteries are cleared and the heart muscle, damaged years earlier, is repaired.

Finally, the molecular machinery of the cells has been restored to working order, and coarser repairs have corrected damaged patterns of cells to restore tissues and organs to a healthy condition. The scaffolding is then removed from the cells, together with most of the temporary cross-links and much of the repair machinery. Most of each cell's active molecules remain blocked, though, to prevent premature, unbalanced activity.

Outside the body, the repair system has grown fresh blood from the patient's own cells. It now transfuses this blood to refill the circulatory system, and acts as a temporary artificial heart. The remaining devices in each cell now adjust the concentration of salts, sugars, ATP, and other small molecules, largely by selectively unblocking each cell's own nanomachinery. With further unblocking, metabolism resumes step by step; the heart muscle is finally unblocked on the verge of contraction. Heartbeat resumes, and the patient emerges into a state of anesthesia. While the attending physicians check that all is going well, the repair system closes the opening in the chest, joining tissue to tissue without a stitch or a scar. The remaining devices in the cells disassemble one another into harmless waste or nutrient molecules. As the patient moves into ordinary sleep, certain visitors enter the room, as long planned.

At last, the sleeper wakes refreshed to the light of a new day—and to the sight of old friends.

MIND, BODY, AND SOUL

Before considering resuscitation, however, some may ask what becomes of the soul of a person in biostasis. Some people would answer that the soul and the mind are aspects of the same thing, of a pattern embodied in the substance of the brain, active during active life and quiescent in biostasis. Assume, though, that the pattern of mind, memory, and personality leaves the body at death, carried by some subtle substance. The possibilities then seem fairly clear. Death in this case has a meaning other than irreversible damage to the brain, being defined instead by the irreversible departure of the soul. This would make biostasis a pointless but harmless gesture—after all, religious leaders have expressed no concern that mere preservation of the body can somehow imprison a soul. Resuscitation would, in this view, presumably require the cooperation of the soul to succeed. The act of placing patients in biostasis has in fact been accompanied by both Catholic and Jewish ceremonies.

With or without biostasis, cell repair cannot bring immortality.

Physical death, however greatly postponed, will remain inevitable for reasons rooted in the nature of the universe. Biostasis followed by cell repair thus seems to raise no fundamental theological issues. It resembles deep anesthesia followed by life-saving surgery: both procedures interrupt consciousness to prolong life. To speak of "immortality" when the prospect is only long life would be to ignore the facts or to misuse words.

REACTIONS AND ARGUMENTS

The prospect of biostasis seems tailor-made to cause future shock. Most people find today's accelerating change shocking enough when it arrives a bit at a time. But the biostasis option is a present-day consequence of a whole series of future breakthroughs. This prospect naturally upsets the difficult psychological adjustments that people make in dealing with physical decline.

Thus far, I have built the case for cell repair and biostasis on a discussion of the commonplace facts of biology and chemistry. But what do professional biologists think about the basic issues? In particular, do they believe (1) that repair machines will be able to correct the kind of cross-linking damage produced by fixation, and (2) that memory is indeed embodied in a preservable form?

After a discussion of molecular machines and their capabilities—a discussion not touching on medical implications—Dr. Gene Brown, professor of biochemistry and chairman of the department of biology at MIT, gave permission to be quoted as stating that: "Given sufficient time and effort to develop artificial molecular machines and to conduct detailed studies of the molecular biology of the cell, very broad abilities should emerge. Among these could be the ability to separate the proteins (or other biomolecules) in cross-linked structures, and to identify, repair, and replace them." This statement addresses a significant part of the cell repair problem. It was consistently endorsed by a sample of biochemists and molecular biologists at MIT and Harvard after similar discussions.

After a discussion of the brain and the physical nature of memory and personality—again, a discussion not touching on medical implications—Dr. Walle Nauta (Institute Professor of Neuroanatomy at MIT) gave permission to be quoted as stating that: "Based on our present knowledge of the molecular biology of neurons, I think most would agree that the changes produced during the consolidation of long-term memory are reflected in corresponding changes in the number and distribution of different protein molecules in the neurons of the brain." Like Dr. Brown's statement, this addresses a key point regarding the workability of biostasis. It, too, was consistently endorsed by other experts when discussed in a context that insulated the experts from any emotional bias that might result from the medical implications of the statement. Further, since these points relate directly to their specialties, Dr. Brown and Dr. Nauta were appropriate experts to ask.

It seems that the human urge to live will incline many millions of people toward using biostasis (as a last resort) if they consider it workable. As molecular technology advances, understanding of cell repair will spread through the popular culture. Expert opinion will increasingly support the idea. Biostasis will grow more common, and its costs will fall. It seems likely that many people will eventually consider biostasis to be the norm, to be a standard lifesaving treatment for patients who have expired.

But until cell repair machines are demonstrated, the all too human tendency to ignore what we haven't seen will slow the acceptance of biostasis. Millions will no doubt pass from expiration to irreversible dissolution because of habit and tradition, supported by weak arguments. The importance of clear foresight in this matter makes it important to consider possible arguments before leaving the topic of life extension and moving on to other matters. Why, then, might biostasis *not* seem a natural, obvious idea?

Because cell repair machines aren't here yet.

It may seem strange to save a person from dissolution in the expectation of restoring health, since the repair technology doesn't exist yet. But is this so much stranger than saving money to put a child through college? After all, the college student doesn't exist yet, either. Saving money makes sense because the child will mature; saving a person makes sense because molecular technology will mature.

We expect a child to mature because we have seen many children mature; we can expect this technology to mature because we have seen many technologies mature. True, some children suffer from congenital shortcomings, as do some technologies, but experts often can estimate the potential of children or technologies while they remain young.

Microelectronic technology started with a few spots and wires on a chip of silicon, but grew into computers on chips. Physicists such as Richard Feynman saw, in part, how far it would lead.

Nuclear technology started with a few atoms splitting in the laboratory under neutron bombardment, but grew into billion-watt reactors and nuclear bombs. Leo Szilard saw, in part, how far it would lead.

Liquid rocket technology started with crude rockets launched from a Massachusetts field, but grew into Moonships and space shuttles. Robert Goddard saw, in part, how far it would lead.

Molecular engineering has started with ordinary chemistry and molecular machines borrowed from cells, but it, too, will grow mighty. It, too, has discernible consequences.

Because tiny machines lack drama.

We tend to expect dramatic results only from dramatic causes, but the world often fails to cooperate. Nature delivers both triumph and disaster in brown paper wrappers.

DULL FACT: Certain electric switches can turn one another on and off. These switches can be made very small, and frugal of electricity.

THE DRAMATIC CONSEQUENCE: When properly connected, these switches form computers, the engines of the information revolution.

DULL FACT: Ether is not too poisonous, yet temporarily interferes with the activity of the brain.

THE DRAMATIC CONSEQUENCE: An end to the agony of surgery on conscious patients, opening a new era in medicine.

DULL FACT: Molds and bacteria compete for food, so some molds have evolved to secrete poisons that kill bacteria.

THE DRAMATIC CONSEQUENCE: Penicillin, the conquest of many bacterial diseases, and the saving of millions of lives.

DULL FACT: Molecular machines can be used to handle molecules and build mechanical switches of molecular size.

THE DRAMATIC CONSEQUENCE: Computer-directed cell repair machines, bringing cures for virtually all diseases.

DULL FACT: Memory and personality are embodied in preservable brain structures.

THE DRAMATIC CONSEQUENCE: Present techniques can prevent dissolution, letting the present generation take advantage of tomorrow's cell repair machines.

In fact, molecular machines aren't even so dull. Since tissues are made of atoms, one should expect a technology able to handle and rearrange atoms to have dramatic medical consequences.

Because this seems too incredible.

We live in a century of the incredible.

In an article entitled "The Idea of Progress" in *Astronautics and Aeronautics,* aerospace engineer Robert T. Jones wrote: "In 1910, the year I was born, my father was a prosecuting attorney. He traveled all the dirt roads in Macon County in a buggy behind a single horse. Last year I flew nonstop from London to San Francisco over the polar regions, pulled through the air by engines of 50,000 horsepower." In his father's day, such aircraft lay at the fringe of science fiction, too incredible to consider.

In an article entitled "Basic Medical Research: A Long-Term Investment" in MIT's *Technology Review,* Dr. Lewis Thomas wrote: "Forty years ago, just before the profession underwent transformation from an art to science and technology, it was taken for granted that the medicine we were being taught was precisely the medicine that would be with us for most of our lives. If anyone had tried to tell us that the power to control bacterial infections was just around the corner, that open-heart surgery or kidney transplants would be possible within a couple of decades, that some kinds of cancer could be cured by chemotherapy, and that we would soon be within reach of a comprehensive, biochemical explanation for genetics and genetically determined diseases, we would have reacted in blank disbelief. We had no reason to believe that medicine would ever change. . . . What this recollection suggests is that we should keep our minds wide open in the future."

Because this sounds too good to be true.

News of a way to avoid the fatality of most fatal diseases may indeed sound too good to be true—as it should, since it is but a small part of a more balanced story. In fact, the dangers of molecular technology roughly balance its promise. In Part Three I will outline reasons for considering nanotechnology more dangerous than nuclear weapons.

Fundamentally, though, nature cares nothing for our sense of good and bad and nothing for our sense of balance. In particular, nature does not hate human beings enough to stack the deck against us. Ancient horrors have vanished before.

Years ago, surgeons strove to amputate legs *fast*. Robert Liston of Edinburgh, Scotland, once sawed through a patient's thigh in a record thirty-three seconds, removing three of his assistant's fingers in the process. Surgeons worked fast to shorten their patients' agony, because their patients remained conscious.

If terminal illness without biostasis is a nightmare today, consider surgery without anesthesia in the days of our ancestors: the knife slicing through flesh, the blood flowing, the saw grating on the bone of a conscious patient. . . . Yet in October of 1846, W. T. G. Morton and J. C. Warren removed a tumor from a patient under ether anesthesia; Arthur Slater states that their success "was rightly hailed as the great discovery of the age." With simple techniques based on a known chemical, the waking nightmare of knife and saw at long last was ended.

With agony ended, surgery increased, and with it surgical infection and the horror of routine death from flesh rotting in the body. Yet in 1867 Joseph Lister published the results of his experiments with phenol, establishing the principles of antiseptic surgery. With simple techniques based on a known chemical, the nightmare of rotting alive shrank dramatically.

Then came sulfa drugs and penicillin, which ended many deadly diseases in a single blow . . . the list goes on.

Dramatic medical breakthroughs have come before, sometimes from new uses of known chemicals, as in anesthesia and antiseptic surgery. Though these advances may have seemed too good to be true, they were true nonetheless. Saving lives by using known chemicals and procedures to produce biostasis can likewise be true.

Because doctors don't use biostasis today.

Robert Ettinger proposed a biostasis technique in 1962. He states that Professor Jean Rostand had proposed the same approach years earlier, and had predicted its eventual use in medicine. Why did biostasis by freezing fail to become popular? In part because of its initial expense, in part because of human inertia, and in part because means for repairing cells remained obscure. Yet the ingrained conservatism of the medical profession has also played a role. Consider again the history of anesthesia.

In 1846, Morton and Warren amazed the world with the "discovery of the age," ether anesthesia. Yet two years earlier, Horace Wells had used nitrous oxide anesthesia, and two years before that, Crawford W. Long had performed an operation using ether. In 1824, Henry Hickman had successfully anesthetized animals using ordinary carbon dioxide; he later spent years urging surgeons in England and France to test nitrous oxide as an anesthetic. In 1799, *a full forty-seven years before the great "discovery,"* and years before Liston's assistant lost his fingers, Sir Humphry Davy wrote: "As nitrous oxide in its extensive operation appears capable of destroying physical pain, it may possibly be used during surgical operations."

Yet as late as 1839 the conquest of pain still seemed an impossible dream to many physicians. Dr. Alfred Velpeau stated: "The abolishment of pain in surgery is a chimera. It is absurd to go on seeking it today. 'Knife' and 'pain' are two words in surgery that must forever be associated in the consciousness of the patient. To this compulsory combination we shall have to adjust ourselves."

Many feared the pain of surgery more than death itself. Perhaps the time has come to awaken from the final medical nightmare.

Because it hasn't been proved to work.

It is true that no experiment can now demonstrate the resuscitation of a patient in biostasis. But a demand for such a demonstration would carry the hidden assumption that modern medicine has neared the final limits of the possible, that it will never be humbled by the achievements of the future. Such a demand might sound cautious and reasonable, but in fact it would smack of overwhelming arrogance.

Unfortunately, a demonstration is exactly what physicians have been trained to request, and for good reason: they wish to avoid useless procedures that may do harm. Perhaps it will suffice that neglect of biostasis leads to obvious and irreversible harm.

TIME, COST, AND HUMAN ACTION

Whether people choose to use biostasis will depend on whether they see it as worth the gamble. This gamble involves the value of life (which is a personal matter), the cost of biostasis (which seems reasonable by the standards of modern medicine), the odds that the technology will work (which seem excellent), and the odds that humanity will survive, develop the technology, and revive people. This final point accounts for most of the overall uncertainty.

Assume that human beings and free societies will indeed survive. (No one can calculate the odds of this, but to assume failure would discourage the very efforts that will promote success.) If so, then technology will continue to advance. Developing assemblers will take years. Studying cells and learning to repair the tissues of patients in biostasis will take still longer. At a guess, developing repair systems and adapting them to resuscitation will take three to ten decades, though advances in automated engineering may speed the process.

The time required seems unimportant, however. Most resuscitated patients will care more about the conditions of life—including the presence of their friends and family—than they will care about the date on the calendar. With abundant resources, the physical conditions of life could be very good indeed. The presence of companions is another matter.

In a recently published survey, over half of those responding said that they would like to live for at least five hundred years, if given a free choice. Informal surveys show that most people would prefer biostasis to dissolution, if they could regain good health and explore a new future with old companions. A few people say that they "want to go when their time comes," but they generally agree that, so long as they can choose further life, their time has not yet come. It seems that many people today share Benjamin Franklin's desire, but in a century able to satisfy it. If biostasis catches on fast enough (or if other life-extension technologies advance fast enough), then a resuscitated patient will awake not to a world of strangers, but to the smiles of familiar faces.

But will people in biostasis be resuscitated? Techniques for placing patients in biostasis are already known, and the costs could become low, at least compared to the costs of major surgery or prolonged hospital care. Resuscitation technology, though, will be complex and expensive to develop. Will people in the future bother?

It seems likely that they will. They may not develop nanotechnology with medicine in mind—but if not, then they will surely develop it to build better computers. They may not develop cell repair machines with resuscitation in mind, but they will surely do so to heal themselves. They may not program repair machines for resuscitation as an act of impersonal charity, but they will have time, wealth, and automated engineering systems, and some of them will have loved ones waiting in biostasis. Resuscitation techniques seem sure to be developed.

With replicators and space resources, a time will come when people have wealth and living space over a thousandfold greater than we have today. Resuscitation itself will require little energy and material even by today's standards. Thus, people contemplating resuscitation will find little conflict between their self-interest and their humanitarian concerns. Common human motives seem enough to ensure that the active population of the future will awaken those in biostasis.

The first generation that will regain youth without being forced to resort to biostasis may well be with us today. The prospect of biostasis simply gives more people more reason to expect long life—it offers an opportunity for the old and a form of insurance for the young. As advances in biotechnology lead toward protein design, assemblers, and cell repair, and as the implications sink in, the expectation of long life will spread. By broadening the path to long life, the biostasis option will encourage a more lively interest in the future. And this will spur efforts to prepare for the dangers ahead.

10

The Limits to Growth

The chess board is the world, the pieces are the phenomena of the universe, the rules of the game are what we call the laws of nature.

—T. H. HUXLEY

IN THE LAST CENTURY we have developed aircraft, spacecraft, nuclear power, and computers. In the next we will develop assemblers, replicators, automated engineering, cheap spaceflight, cell repair machines, and much more. This series of breakthroughs may suggest that the technology race will advance without limit. In this view, we will break through all conceivable barriers, rushing off into the infinite unknown—but this view seems false.

The laws of nature and the conditions of the world will limit what we do. Without limits, the future would be wholly unknown, a formless thing making a mockery of our efforts to think and plan. With limits, the future is still a turbulent uncertainty, but it is forced to flow within certain bounds.

From natural limits, we learn something about the problems and opportunities we face. Limits define the boundaries of the possible, telling us what resources we can use, how fast our spacecraft will fly, and what our nanomachines will and won't be able to do.

Discussing limits is risky: we can be more sure that something *is* possible than that it *isn't*. Engineers can make do with approximations and special cases. And given tools, materials, and time, they can demonstrate possibilities directly. Even when doing exploratory design, they can stay well within the realm of the possible by staying well away from the limits. Scientists, in contrast, cannot prove a general theory—and every general claim of impossibility is itself a sort of general theory. No specific experiment (someplace, sometime) can prove something to be impossible (everywhere, forever). Neither can any number of specific experiments.

Still, general scientific laws do describe limits to the possible. Although scientists cannot *prove* a general law, they have evolved our best available picture of how the universe works. And even if exotic experiments and elegant mathematics again transform our concept of physical law, few engineering limits will budge. Relativity didn't affect automobile designs.

The mere existence of ultimate limits doesn't mean that they are about to choke us, yet many people have been drawn to the idea that limits will end growth soon. This notion simplifies their picture of the future by leaving out the strange new developments that growth will bring. Other people favor the vaguer notion of limitless growth —a notion that blurs their picture of the future by suggesting that it will be utterly incomprehensible.

People who confuse science with technology tend to become confused about limits. As software engineer Mark S. Miller points out, they imagine that new knowledge always means new know-how; some even imagine that knowing *everything* would let us do *anything.* Advances in technology do indeed bring new know-how, opening new possibilities. But advances in basic science simply redraw our map of ultimate limits; this often shows new *im*possibilities. Einstein's discoveries, for example, showed that nothing can catch up with a fleeing light ray.

THE STRUCTURE OF THE VACUUM

Is the speed of light a real limit? People once spoke of a "sound barrier" that some believed would stop an airplane from passing the speed of sound. Then at Edwards Air Force Base in 1947, Chuck Yeager split the October sky with a sonic boom. Today, some people speak of a "light barrier," and ask whether it, too, may fall.

Unfortunately for science fiction writers, this parallel is superficial. No one could ever maintain that the speed of sound was a true physical limit. Meteors and bullets exceeded it daily, and even cracking whips cracked the "sound barrier." But no one has seen anything move faster than light. Distant spots seen by radio telescopes sometimes *appear* to move faster, but simple tricks of perspective easily explain how this can be. Hypothetical particles called "tachyons" would move faster than light, if they were to exist—but none has been found, and current theory doesn't predict them. Experimenters have pushed protons to over 99.9995 percent of the speed of light, with results that match Einstein's predictions perfectly. When pushed ever faster, a particle's speed creeps closer to the speed of light, while its energy (mass) grows without bound.

On Earth, a person can walk or sail only to a certain distance, but no mysterious edge or barrier suddenly blocks travel. The Earth is simply round. The speed limit in space no more implies a "light barrier" than the distance limit on Earth implies a wall. Space itself—the vacuum that holds all energy and matter—has properties. One of these is its geometry, which can be described by regarding time as a special dimension. This geometry makes the speed of light recede before an accelerating spaceship much as the horizon recedes before a moving sea ship: the speed of light, like the horizon, is always equally remote in all directions. But the analogy dies here—this similarity has nothing to do with the curvature of space. It is enough to remember that the limiting speed is nothing so crude or so breakable as a "light barrier." Objects can always go faster, just no faster than light.

People have long dreamed of gravity control. In the 1962 edition of *Profiles of the Future*, Arthur C. Clarke wrote that "Of all the forces, gravity is the most mysterious and the most implacable," and then went on to suggest that we will someday develop convenient devices for controlling gravity. Yet is gravity really so mysterious? In the general theory of relativity, Einstein described gravity as curvature in the space-time structure of the vacuum. The mathematics describing this is elegant and precise, and it makes predictions that have passed every test yet contrived.

Gravity is neither more nor less implacable than other forces. No one can make a boulder lose its gravity, but neither can anyone make an electron lose its electric charge or a current its magnetic field. We control electric and magnetic fields by moving the particles that create them; we can control gravitational fields similarly, by moving ordinary masses. It seems that we cannot learn the secret of gravity control because we already have it.

A child with a small magnet can lift a nail, using a magnetic field to overwhelm Earth's gravitational pull. But unfortunately for eager gravitational engineers, using gravity to lift a nail requires a tremendous mass. Hanging Venus just over your head would almost do the job—until it fell on you.

Engineers stir up electromagnetic waves by shaking electric charges back and forth in an antenna; one can stir up gravity waves by shaking a rock in the air. But again, the gravitational effect is weak. Though a one-kilowatt radio station is nothing extraordinary, all the shaking and spinning of all the masses in the solar system put together fails to radiate as much as a kilowatt of gravity waves.

We understand gravity well enough; it simply isn't much use in building machines much lighter than the Moon. But devices using large masses do work. A hydroelectric dam is part of a gravity machine (the other part being the Earth) that extracts energy from falling mass. Machines using black holes will be able to extract energy from falling mass with over fifty percent efficiency, based on $E = mc^2$. Lowering a single bucket of water into a black hole would yield as much energy as pouring several trillion buckets of water through the generators of a kilometer-high dam.

Because the laws of gravity describe how the vacuum curves, they also apply to science-fiction style "space warps." It seems that tunnels from one point in space to another would be unstable, even if they could be created in the first place. This prevents future spaceships from reaching distant points faster than light by taking a shortcut around the intervening space, and this limit to travel in turn sets a limit to growth.

Einstein's laws seem to give an accurate description of the overall geometry of the vacuum. If so, then the limits that result seem inescapable: you can get rid of almost anything, but not the vacuum itself.

Other laws and limits seem inescapable for similar reasons. In fact, physicists have increasingly come to regard all of physical law in terms of the structure of the vacuum. Gravity waves are a certain kind of ripple in the vacuum; black holes are a certain kind of kink. Likewise, radio waves are another kind of ripple in the vacuum, and elementary particles are other, very different kinds of kinks (which in some theories resemble tiny, vibrating strings). In this view, there is only one substance in the universe—the vacuum—but one that takes on a variety of forms, including those patterns of particles that we call "solid matter." This view suggests the inescapable quality of natural law. If a single substance fills the universe, *is* the universe, then its properties limit all that we can do.

The strangeness of modern physics, however, leads many people to distrust it. The revolutions that brought quantum mechanics and relativity gave rise to talk of "the uncertainty principle," "the wave nature of matter," "matter being energy," and "curved space-time." An air of paradox surrounds these ideas and thus physics itself. It is understandable that new technologies should seem odd to us, but why should the ancient and immutable laws of nature turn out to be bizarre and shocking?

Our brains and languages evolved to deal with things vastly larger than atoms, moving at a tiny fraction of the speed of light. They do a tolerable job of this, though it took people centuries to learn to describe the motion of a falling rock. But we have now stretched our knowledge far beyond the ancient world of the senses. We have found things (matter waves, curved space) that seem bizarre—and some that are simply beyond our ability to visualize. But "bizarre" does not mean mysterious and unpredictable. Mathematics and experiments still work, letting scientists vary and select theories, evolving them to fit even a peculiar reality. Human minds have proved remarkably flexible, but it is no great surprise to find that we cannot always visualize the invisible.

Part of the reason that physics seems so strange is that people crave oddities, and tend to spread memes that describe things as odd. Some people favor ideas that coat the world in layers of ghost stuff and fill it with grade-B mysteries. Naturally, they favor and spread memes that make matter seem immaterial and quantum mechanics seem like a branch of psychology.

Relativity, it is said, reveals that *matter* (that plain old stuff that people think they understand) is really *energy* (that subtle, mysterious stuff that makes things happen). This encourages a smiling vagueness about the mystery of the universe. It might be clearer to say that relativity reveals *energy* to be a form of *matter:* in all its forms, energy has mass. Indeed, lightsails work on this principle, through the impact of mass on a surface. Light even comes packaged in particles.

Consider also the Heisenberg uncertainty principle, and the related fact that "the observer always affects the observed." The uncertainty principle is intrinsic to the mathematics describing ordinary matter (giving atoms their very size), but the associated "effect of the observer" has been presented in some popular books as a magical influence of consciousness on the world. In fact, the core idea is more prosaic. Imagine looking at a dust mote in a light beam: When you observe the reflected light, you certainly affect it—your eye absorbs it. Likewise, the light (with its mass) affects the dust mote: it bounces off the mote, exerting a force. The result is not an effect of your mind on the dust, but of the light on the dust. Though quantum measurement has peculiarities far more subtle than this, none involves the mind reaching out to change reality.

Finally, consider the "twin paradox." Relativity predicts that, if one of a pair of twins flies to another star and back at near the speed of light, the traveling twin will be younger than the stay-at-home twin. Indeed, measurements with accurate clocks demonstrate the time-slowing effect of rapid motion. But this is not a "paradox"; it is a simple fact of nature.

WILL PHYSICS AGAIN BE UPENDED?

In 1894 the eminent physicist Albert A. Michelson stated: "The more important fundamental laws and facts of physical science have all been discovered, and these are now so firmly established that the possibility of their ever being supplanted in consequence of new discoveries is exceedingly remote . . . Our future discoveries must be looked for in the sixth place of decimals."

But in 1895, Roentgen discovered X rays. In 1896, Becquerel discovered radioactivity. In 1897, Thomson discovered the electron. In 1905, Einstein formulated the special theory of relativity (and thus explained Michelson's own 1887 observations regarding the speed of light). In 1905, Einstein also presented the photon theory of light. In 1911, Rutherford discovered the atomic nucleus. In 1915, Einstein formulated the general theory of relativity. In 1924–30, de Broglie, Heisenberg, Bohr, Pauli, and Dirac developed the foundations of quantum mechanics. In 1929, Hubble announced evidence for the expansion of the universe. In 1931, Michelson died.

Michelson had made a memorable mistake. People still point to his statement and what followed to support the view that we shouldn't (ever?) claim any firm understanding of natural law, or of the limits to the possible. After all, if Michelson was so sure and yet so wrong, shouldn't we fear repeating his mistake? The great revolution in physics has led some people to conclude that science will hold endless important surprises—even surprises important to engineers. But are we likely to encounter such an important upheaval again?

Perhaps not. The content of quantum mechanics was a surprise, yet before it appeared, physics was obviously and grossly incomplete. Before quantum mechanics, you could have walked up to any scientist, grinned maliciously, rapped on his desk and asked, "What holds this thing together? Why is it brown and solid, while air is transparent and gaseous?" Your victim might have said something vague about atoms and their arrangements, but when you pressed for an explanation, you would at best have gotten an answer like "Who knows? Physics can't explain matter yet!" Hindsight is all too easy, yet in a world made of matter, inhabited by material people using material tools, this ignorance of the nature of matter was a gap in human knowledge that Michelson should perhaps have noted. It was a gap not in "the sixth place of decimals" but in the first.

It is also worth observing the extent to which Michelson was *right*. The laws of which he spoke included Newton's laws of gravity and motion, and Maxwell's laws of electromagnetism. And indeed, under conditions common in engineering, these laws *have* been modified only "in the sixth place of decimals." Einstein's laws of gravity and motion match Newton's laws closely except under extreme conditions of gravitation and speed; the quantum electrodynamic laws of Feynman, Schwinger, and Tomonaga match Maxwell's laws closely except under extreme conditions of size and energy.

Further revolutions no doubt lurk around the edges of these theories. But those edges seem far from the world of living things and the machines we build. The revolutions of relativity and quantum mechanics changed our knowledge about matter and energy, but matter and energy *themselves* remained unchanged—they are real and care nothing for our theories. Physicists now use a single set of laws to describe how atomic nuclei and electrons interact in atoms, molecules, molecular machines, living things, planets, and stars. These laws are not yet completely general; the quest for a unified theory of all physics continues. But as physicist Stephen W. Hawking states, "at the moment we have a number of partial laws which govern the behavior of the universe under all but the most extreme conditions." And by an engineer's standards, those conditions are *extraordinarily* extreme.

Physicists regularly announce new particles observed in the debris from extremely energetic particle collisions, but you cannot buy one of these new particles in a box. And this is important to recognize, because if a particle cannot be kept, then it cannot serve as a component of a stable machine. Boxes and their contents are made of electrons and nuclei. Nuclei, in turn, are made of protons and neutrons. Hydrogen atoms have single protons as their nuclei; lead atoms have eighty-two protons and over a hundred neutrons. Isolated neutrons disintegrate in a few minutes. Few other stable particles are known: photons—the particles of light—are useful and can be trapped for a time; neutrinos are almost undetectable and cannot even be trapped. These particles (save the photon) have matching antiparticles. *All other known particles disintegrate in a few millionths of a second or less.* Thus, the only known building blocks for hardware are electrons and nuclei (or their antiparticles, for special *isolated* applications); these building blocks ordinarily combine to form atoms and molecules.

Yet despite the power of modern physics, our knowledge still has obvious holes. The shaky state of elementary particle theory leaves some limits uncertain. We may find new stable, "boxable" particles, such as magnetic monopoles or free quarks; if so, they will doubtless have uses. We may even find a new long-range field or form of radiation, though this seems increasingly unlikely. Finally, some new way of smashing particles together may improve our ability to convert known particles into other known particles.

But in general, complex hardware will require complex, stable patterns of particles. Outside the environment of a collapsed star, this means patterns of atoms, which are well described by relativistic quantum mechanics. The frontiers of physics have moved on. On a theoretical level, physicists seek a unified description of the interactions of all possible particles, even the most short-lived particles. On an experimental level, they study the patterns of subatomic debris created by high-energy collisions in particle accelerators. So long as no new, stable, and useful particle comes out of such a collision—or turns up as the residue of past cosmic upheavals—atoms will remain the sole building blocks of stable hardware. And engineering will remain a game that is played with already known pieces according to already known rules. New particles would add pieces, not eliminate rules.

THE LIMITS TO HARDWARE

Is molecular machinery really the end of the road where miniaturization is concerned? The idea that molecular machinery might be a step toward a smaller "nuclear machinery" seems natural enough. One young man (a graduate student in economics at Columbia University) heard of molecular technology and its ability to manipulate atoms, and immediately concluded that molecular technology could do almost anything, even turn falling nuclear bombs into harmless lead bricks at a distance.

Molecular technology can do no such thing. Turning plutonium into lead (whether at a distance or not) lies beyond molecular technology for the same reason that turning lead into gold lay beyond an alchemist's chemistry. Molecular forces have little effect on the atomic nucleus. The nucleus holds over 99.9 percent of an atom's mass and occupies about 1/1,000,000,000,000,000 of its volume. Compared to the nucleus, the rest of an atom (an electron cloud) is less than airy fluff. Trying to change a nucleus by poking at it with a molecule is even more futile than trying to flatten a steel ball bearing by waving a ball of cotton candy at it. Molecular technology can sort and rearrange atoms, but it cannot reach into a nucleus to change an atom's type.

Nanomachines may be no help in *building* nuclear-scale machines, but could such machines even *exist?* Apparently not, at least under any conditions we can create in a laboratory. Machines must have a number of parts in close contact, but close-packed nuclei repel each other with ferocity. When splitting nuclei blasted Hiroshima, most of the energy was released by the violent electrostatic repulsion of the freshly split halves. The well-known difficulty of nuclear fusion stems from this same problem of nuclear repulsion.

In addition to splitting or fusing, nuclei can be made to emit or absorb various types of radiation. In one technique, they are made to gyrate in ways that yield useful information, letting doctors make medical images based on nuclear magnetic resonance. But all these phenomena rely only on the properties of well-separated nuclei. Isolated nuclei are too simple to act as machines or electronic circuits. Nuclei can be forced close together, but only under the immense pressures found inside a collapsed star. Doing engineering there would present substantial difficulties, even if a collapsed star were handy.

This returns us to the basic question, What can we accomplish by properly arranging atoms? Some limits already seem clear. The strongest materials possible will have roughly ten times the strength of today's strongest steel wire. (The strongest material for making a cable appears to be carbyne, a form of carbon having atoms arranged in straight chains.) It seems that the vibrations of heat will, at ordinary pressures, tear apart the most refractory solids at temperatures around four thousand degrees Celsius (roughly fifteen hundred degrees cooler than the Sun's surface).

These brute properties of matter—strength and heat resistance—cannot be greatly improved through complex, clever arrangements of atoms. The best arrangements seem likely to be fairly simple and regular. Other fairly simple goals include transmitting heat, insulating against heat, transmitting electricity, insulating against electricity, transmitting light, reflecting light, and absorbing light.

With some goals, pursuit of perfection will lead to simple designs; with others, it will lead to design problems beyond any hope of solving. Designing the best possible switching component for a computer may prove fairly easy; designing the best possible computer will be vastly more complex. Indeed, what we consider to be "the best possible" will depend on many factors, including the costs of matter, energy and time—and on what we want to compute. In any engineering project, what we call "better" depends on indefinitely many factors, including ill-defined and changing human wants. What is more, even where "better" is well defined, the cost of seeking the final increment of improvement that separates the best from the merely excellent may not be worth paying. We can ignore all such issues of complexity and design cost, though, when considering whether limits actually exist.

To define a limit, one must choose a direction, a scale of quality. With that direction defining "better," there will definitely be a best. The arrangement of atoms determines the properties of hardware, and according to quantum mechanics, the number of possible arrangements is finite—more than just astronomically large, yet not infinite. It follows mathematically that, given a clear goal, some one of these arrangements must be best, or tied for best. As in chess, the limited number of pieces and spaces limits the arrangements and hence the possibilities. In both chess and engineering, though, the variety possible within those limits is inexhaustible.

Just knowing the fundamental laws of matter isn't enough to tell us exactly where all the limits lie. We still must face the complexities of design. Our knowledge of some limits remains loose: "We know only that the limit lies between here (a few paces away) and there (that spot near the horizon)." Assemblers will open the way to the limits, wherever they are, and automated engineering systems will speed progress along the road. The absolute best will often prove elusive, but the runners-up will often be nearly as good.

As we approach genuine limits, our abilities will, in ever more areas of technology, cease growing. Advances in these fields will stop not merely for a decade or a century, but permanently.

Some may balk at the word "permanently," thinking "No improvements in a thousand years? In a *million* years? This must be an overstatement." Yet where we reach true physical limits, we will go no further. The rules of the game are built into the structure of the vacuum, into the structure of the universe. No rearranging of atoms, no clashing of particles, no legislation or chanting or stomping will move natural limits one whit. We may misjudge the limits today, but wherever the real limits lie, there they will remain.

This look at natural law shows limits to the *quality* of things. But we also face limits to *quantity,* set not only by natural law but by the way that matter and energy are arranged in the universe as we happen to find it. The authors of *The Limits to Growth,* like so many others, attempted to describe these limits without first examining the limits to technology. This gave misleading results.

ENTROPY: A LIMIT TO ENERGY USE

Recently, some authors have described the accumulation of waste heat and disorder as ultimate limits to human activity. In *The Lean Years—Politics in the Age of Scarcity,* Richard Barnet writes:

"It is ironical that the rediscovery of limits coincides with two of the most audacious technological feats in human history. One is genetic engineering, the sudden glimpse of a power to shape the very stuff of life. The other is the colonization of space. These breakthroughs encourage fantasies of power, but they do not break the ecological straightjacket known as the Second Law of Thermodynamics: Ever greater consumption of energy produces ever greater quantities of heat, which never disappear, but must be counted as a permanent energy cost. Since accumulation of heat can cause ecological catastrophe, these costs limit man's adventure in space as surely as on earth."

Jeremy Rifkin (with Ted Howard) has written an entire book on thermodynamic limits and the future of humanity, titled *Entropy: A New World View.*

Entropy is a standard scientific measure of waste heat and disorder. Whenever activities consume useful energy, they produce entropy; the entropy of the world therefore increases steadily and irreversibly. Ultimately, the dissipation of useful energy will destroy the basis of life. As Rifkin says, this idea may seem too depressing to consider, but he argues that we must face the terrible facts about entropy, humanity, and the Earth. But are these facts so terrible?

Barnet writes that accumulating heat is a permanent energy cost, limiting human action. Rifkin states that "pollution is the sum total of all of the available energy in the world that has been transformed into unavailable energy." This unavailable energy is chiefly low-temperature waste heat, the sort that makes television sets get warm. But does heat really accumulate, as Barnet fears? If so, then the Earth must be growing steadily hotter, minute by minute and year by year. We should be roasting now, if our ancestors weren't frozen solid. Somehow, though, continents manage to get cold at night, and colder yet during the winter. During ice ages, the whole Earth cools off.

Rifkin takes another tack. He states that "the fixed endowment of terrestrial matter that makes up the earth's crust is constantly dissipating. Mountains are wearing down and topsoil is being blown away with each passing second." By "blown away" Rifkin doesn't mean blown into space or blown out of existence; he just means that the mountain's atoms have become jumbled together with others. Yet this process, he argues, means our doom. The jumbling of atoms makes them "unavailable matter," as a consequence of the "fourth law of thermodynamics," propounded by economist Nicholas Georgescu-Roegen: "In a closed system, the material entropy must ultimately reach a maximum," or (equivalently) that "unavailable matter cannot be recycled." Rifkin declares that the Earth is a closed system, exchanging energy but not matter with its surroundings, and that therefore "here on earth material entropy is continually increasing and must ultimately reach a maximum," making Earth's life falter and die.

A grim situation indeed—the Earth has been degenerating for billions of years. Surely the end must be near!

But can this really be true? As life developed, it brought more order to Earth, not less; the formation of ore deposits did the same. The idea that Earth has degenerated seems peculiar at best (but then, Rifkin thinks evolution is bunk). Besides, since matter and energy are essentially the same, how can a valid law single out something called "material entropy" in the first place?

Rifkin presents perfume spreading from a bottle into the air in a room as an example of "dissipating matter," of material entropy increasing—of matter becoming "unavailable." The spread of salt into water in a bottle will serve equally well. Consider, then, a test of the "fourth law of thermodynamics" in the Salt-Water Bottle Experiment:

Imagine a bottle having a bottom with a partition, dividing it into two basins. In one sits salt, in the other sits water. A cork plugs the bottle's neck: this closes the system and makes the so-called fourth law of thermodynamics apply. The bottle's contents are in an organized state: their material entropy is not at a maximum—yet.

Now, pick up the bottle and shake it. Slosh the water into the other basin, swirl it around, dissolve the salt, increase the entropy—go wild! In such a closed system, the "fourth law of thermodynamics" says that this increase in the material entropy should be permanent. All of Rifkin's alarums about the steady, inevitable increase of Earth's entropy rest on this principle.

To see if there is any basis for Rifkin's new worldview, take the bottle and tip it, draining the salty water into one basin. This should make no difference, since the system remains closed. Now set the bottle upright, placing the saltwater side in sunlight and the empty side in shade. Light shines in and heat leaks out, but the system remains as closed as the Earth itself. But watch—the sunlight evaporates water, which condenses in the shade! Fresh water slowly fills the empty basin, *leaving the salt behind.*

Rifkin himself states that "in science, only one uncompromising exception is enough to invalidate a law." This thought experiment, which mimics how natural salt deposits have formed on Earth, invalidates the law on which he founds his whole book. So do plants. Sunlight brings energy from space; heat radiated back into space carries away entropy (of which there is only one kind). Therefore, entropy can decline in a closed system and flowers can bloom on Earth for age upon age.

Rifkin is right in saying that "it's possible to reverse the entropy process in an isolated time and place, but only by using up energy in the process and thus increasing the overall entropy of the environment." But both Rifkin and Barnet make the same mistake: when they write of the environment, they imply the Earth—but the law applies to the environment as a whole, and that whole is the *universe*. In effect, Rifkin and Barnet ignore both the light of the Sun and the cold black of the night sky.

According to Rifkin, his ideas destroy the notion of history as progress, transcending the modern worldview. He calls for sacrifice, stating that "no Third World nation should harbor hopes that it can ever reach the material abundance that has existed in America." He fears panic and bloodshed. Rifkin finishes by informing us that "the Entropy Law answers the central question that every culture throughout history has grappled with: How should human beings behave in the world?" His answer? "The ultimate moral imperative, then, is to waste as little energy as possible."

This would seem to mean that we must *save* as much energy as possible, seeking to eliminate waste. But what is the greatest nearby energy waster? Why, the Sun, of course—it wastes energy trillions of times faster than we humans do. If taken seriously, it seems that Rifkin's ultimate moral imperative therefore urges: "Put out the Sun!"

This silly consequence should have tipped Rifkin off. He and many others hold views that smack of a pre-Copernican arrogance: they presume that the Earth is the whole world and that what people do is automatically of cosmic importance.

There is a genuine entropy law, of course: the second law of thermodynamics. Unlike the bogus "fourth law," it is described in textbooks and used by engineers. It will indeed limit what we do. Human activity will generate heat, and Earth's limited ability to radiate heat will set a firm limit to the amount of Earth-based industrial activity. Likewise, we will need winglike panels to radiate waste heat from our starships. Finally, the entropy law will—at the far end of an immensity of time—bring the downfall of the universe as we know it, limiting the lifespan of life itself.

Why flog the carcass of Rifkin's *Entropy?* Simply because today's information systems often present even stillborn ideas as if they were alive. By encouraging false hopes, false fears, and misguided action, these ideas can waste the efforts of people actively concerned about long-range world problems.

Among those whose praise appears on the back cover of Rifkin's book ("an inspiring work," "brilliant work," "earthshaking," "should be taken to heart") are a Princeton professor, a talk-show host, and two United States senators. A seminar at MIT ("The Finite Earth—World Views for a Sustainable Future") featured Rifkin's book.

All the seminar's sponsors were from nontechnical departments. Most senators in our technological society lack training in technology, as do most professors and talk-show hosts. Georgescu-Roegen himself, inventor of the "fourth law of thermodynamics," has extensive credentials—as a social scientist.

The entropy threat is an example of blatant nonsense, yet its inventors and promoters aren't laughed off the public stage. Imagine a thousand, a million similar distortions—some subtle, some brazen, but all warping the public's understanding of the world. Now imagine a group of democratic nations suffering from an infestation of such memes while attempting to cope with an era of accelerating technological revolution. We have a real problem. To make our survival more likely, we will need better ways to weed our memes, to make room for sound understanding to grow. In Chapters 13 and 14 I will report on two proposals for how we might do this.

THE LIMITS TO RESOURCES

Natural law limits the quality of technology, but within these limits we will use replicating assemblers to produce superior spacecraft. With them, we will open space wide and deep.

Today Earth has begun to seem small, arousing concerns that we may deplete its resources. Yet the energy we use totals less than 1/10,000 of the solar energy striking Earth; we worry not about the supply of energy as such, but about the supply of convenient gas and oil. Our mines barely scratch the surface of the globe; we worry not about the sheer quantity of resources, but about their convenience and cost. When we develop pollution-free nanomachines to gather solar energy and resources, Earth will be able to support a civilization far larger and wealthier than any yet seen, yet suffer less harm than we inflict today. The potential of Earth makes the resources we now use seem insignificant by comparison.

Yet Earth is but a speck. The asteroidal debris left over from the formation of the planets will provide materials enough to build a thousand times Earth's land area. The Sun floods the solar system with a billion times the power that reaches Earth. The resources of the solar system are truly vast, making the resources of Earth seem insignificant by comparison.

Yet the solar system is but a speck. The stars that crowd the night sky are suns, and the human eye can see only the closest. Our galaxy holds a hundred billion suns, and many no doubt pour their light on dead planets and asteroids awaiting the touch of life. The resources of our galaxy make even our solar system seem insignificant by comparison.

Yet our galaxy is but a speck. Light older than our species shows galaxies beyond ours. The visible universe holds a hundred billion galaxies, each a swarm of billions of suns. The resources of the visible universe make even our galaxy seem insignificant by comparison.

With this we reach the limits of knowledge, if not of resources. The solar system seems answer enough to Earth's limits—and if the rest of the universe remains unclaimed by others, then our prospects for expansion boggle the mind several times over. Does this mean that replicating assemblers and cheap spaceflight will end our resource worries?

In a sense, opening space will burst our limits to growth, since we know of no end to the universe. Nevertheless, Malthus was essentially right.

MALTHUS

In his 1798 *Essay on the Principle of Population,* Thomas Robert Malthus, an English clergyman, presented the ancestor of all modern limits-to-growth arguments. He noted that freely growing populations tend to double periodically, thus expanding exponentially. This makes sense: since all organisms are descended from successful replicators, they tend to replicate when given a chance. For the sake of argument, Malthus assumed that resources—the food supply—could increase by a fixed amount per year (a process called linear growth, since it plots as a line on a graph). Since mathematics shows that *any* fixed rate of exponential growth will eventually outstrip *any* fixed rate of linear growth, Malthus argued that population growth, if unchecked, would eventually outrun food production.

Authors have repeated variations on this idea ever since, in books like *The Population Bomb* and *Famine—1975!*, yet food production has kept pace with population. Outside Africa, it has even pulled ahead. Was Malthus wrong?

Not fundamentally: he was wrong chiefly about timing and details. Growth on Earth does face limits, since Earth has limited room, whether for farming or anything else. Malthus failed to predict when limits would pinch us chiefly because he failed to anticipate breakthroughs in farm equipment, crop genetics, and fertilizers.

Some people now note that exponential growth will overrun the fixed stock of Earth's resources, a simpler argument than the one Malthus made. Though space technology will break this limit, it will not break all limits. Even if the universe were infinitely large, we still could not travel infinitely fast. The laws of nature will limit the rate of growth: Earth's life will spread no faster than light.

Steady expansion will open new resources at a rate that will increase as the frontier spreads deeper and wider into space. This will result not in linear growth, but in cubic growth. Yet Malthus was essentially right: exponential growth will outrun cubic growth as easily as it would linear. Calculations show that unchecked population growth, with or without long life, would overrun available resources in about one or two thousand years at most. Unlimited exponential growth remains a fantasy, even in space.

WILL SOMEONE STOP US?

Do other civilizations already own the resources of the universe? If so, then they would represent a limit to growth. The facts about evolution and technological limits shed useful light on this question.

Since many Sunlike stellar systems are many hundreds of millions of years older than our solar system, some civilizations (if any substantial number exist) should be many hundreds of millions of years ahead of ours. We would expect at least some of these civilizations to do what all known life has done: spread as far as it can. Earth is green not just in the oceans where life began, but on shores, hills, and mountains. Green plants have now spread to stations in orbit; if we prosper, Earth's plants will spread to the stars. Organisms spread as far as they can, then a bit farther. Some fail and die, but the successful survive and spread farther yet. Settlers bound for America sailed and sank, and landed and starved, but some survived to found new nations. Organisms everywhere will feel the pressures that Malthus described, because they will have evolved to survive and spread; genes and memes both push in the same direction. If extraterrestrial civilizations exist, and if even a small fraction were to behave as all life on Earth does, then they should by now have spread across space.

Like us, they would tend to evolve technologies that approach the limits set by natural law. They would learn how to travel near the speed of light, and competition or sheer curiosity would drive some to do so. Indeed, only highly organized, highly stable societies could restrain competitive pressures well enough to *avoid* exploding outward at near the speed of light. By now, after hundreds of millions of years, even widely scattered civilizations would have spread far enough to meet each other, dividing all of space among them.

If these civilizations are indeed everywhere, then they have shown great restraint and hidden themselves well. They would have controlled the resources of whole galaxies for many millions of years, and faced limits to growth on a cosmic scale. An advanced civilization pushing its ecological limits would, almost by definition, not waste both matter and energy. Yet we see such waste in all directions, as far as we can see spiral galaxies: their spiral arms hold dust clouds made of wasted matter, backlit by wasted starlight.

If such advanced civilizations existed, then our solar system would lie in the realm of one of them. If so, then it would now be their move—we could do nothing to threaten them, and they could study us as they pleased, with or without our cooperation. Sensible people would listen if they firmly stated a demand. But if they exist, they must be hiding themselves—and keeping any local laws secret.

The idea that humanity is alone in the visible universe is consistent with what we see in the sky and with what we know about the origin of life. No bashful aliens are needed to explain the facts. Some say that since there are so many stars, there must surely be other civilizations among them. But there are fewer stars in the visible universe than there are molecules in a glass of water. Just as a glass of water need not contain every possible chemical (even downstream from a chemical plant), so other stars need not harbor civilizations.

We know that competing replicators tend to expand toward their ecological limits, and that resources are nonetheless wasted throughout the universe. We have received no envoy from the stars, and we apparently lack even a tolerably humane zookeeper. There may well be no one there. If they do not exist, then we need not consider them in our plans. If they do exist, then they will overrule our plans according to their own inscrutable wishes, and there seems no way to prepare for the possibility. Thus for now, and perhaps forever, we can make plans for our future without concern for limits imposed by other civilizations.

GROWTH WITHIN LIMITS

Whether anyone else is out there or not, we are on our way. Space waits for us, barren rock and sunlight like the barren rock and sunlight of Earth's continents a billion years ago, before life crept forth from the sea. Our engineers are evolving memes that will help us create fine spaceships and settlements: we will settle the land of the solar system in comfort. Beyond the rich inner solar system lies the cometary cloud—a vast growth medium that thins away into the reaches of interstellar space, then thickens once more around other star systems, with fresh suns and sterile rock awaiting the touch of life.

Although endless exponential growth remains a fantasy, the spread of life and civilization faces no fixed bound. Expansion will proceed, if we survive, because we are part of a living system and life tends to spread. Pioneers will move outward into worlds without end. Others will remain behind, building settled cultures throughout the oases of space. In any settlement, the time will come when the frontier lies far away, then farther. For the bulk of the future, most people and their descendants will live with limits to growth.

We may like or dislike limits to growth, but their reality is independent of our wishes. Limits exist wherever goals are clearly defined.

But on frontiers where standards keep changing, this idea of limits becomes irrelevant. In art or mathematics the value of work depends on complex standards, subject to dispute and change. One of those standards is novelty, and this can never be exhausted. Where goals change and complexity rules, limits need not bind us. To the creation of symphony and song, paintings and worlds, software, theorems, films, and delights yet unimagined, there seems no end. New technologies will nurture new arts, and new arts will bring new standards.

The world of brute matter offers room for great but limited growth. The world of mind and pattern, though, holds room for endless evolution and change. The possible seems room enough.

VIEWS OF LIMITS

The idea of great advances within firm limits isn't evolved to feel pleasing, but to be accurate. Limits outline possibilities, and some may be ugly or terrifying. We need to prepare for the breakthroughs ahead, yet many futurists studiously pretend that no breakthroughs will occur.

This school of thought is associated with *The Limits to Growth*, published as a report to the Club of Rome. Professor Mihajlo D. Mesarović later coauthored *Mankind at the Turning Point*, published as the second report to the Club of Rome. Professor Mesarović develops computer models like the one used in *The Limits to Growth*— each is a set of numbers and equations that purports to describe future changes in the world's population, economy, and environment. In the spring of 1981, he visited MIT to address "The Finite Earth: Worldviews for a Sustainable Future," the same seminar that featured Jeremy Rifkin's *Entropy*. He described a model intended to give a rough description of the next century. When asked whether he or any of his colleagues had allowed for even one future breakthrough comparable to, say, the petroleum industry, aircraft, automobiles, electric power, or computers—perhaps self-replicating robotic systems or cheap space transportation?—he answered directly: "No."

Such models of the future are obviously bankrupt. Yet some people seem willing—even eager—to believe that breakthroughs will suddenly cease, that a global technology race that has been gaining momentum for centuries will screech to a halt in the immediate future.

The habit of neglecting or denying the possibility of technological advance is a common problem. Some people believe in snugly fitting limits because they have heard respected people spin plausible-sounding arguments for them. Yet it seems that some people must be responding more to wish than to fact, after this century of accelerating advance. Snug limits would simplify our future, making it easier to understand and more comfortable to think about. A belief in snug limits also relieves a person of certain concerns and responsibilities. After all, if natural forces will halt the technology race in a convenient and automatic fashion, then we needn't try to understand and control it.

Best of all, this escape doesn't *feel* like escapism. To contemplate visions of global decline must give the feeling of facing harsh facts without flinching. Yet such a future would be nothing really new: it would force us toward the familiar miseries of the European past or the Third World present. Genuine courage requires facing reality, facing accelerating change in a world that has no automatic brakes. This poses intellectual, moral, and political challenges of greater substance.

Warnings of bogus limits do double harm. First, they discredit the very idea of limits, blunting an intellectual tool that we will need to understand our future. But worse, such warnings distract attention from our real problems. In the Western world there is a lively political tradition that fosters suspicion of technology. To the extent that it first disciplines its suspicions by testing them against reality and then chooses workable strategies for guiding change, this tradition can contribute mightily to the survival of life and civilization. But people concerned about technology and the future are a limited resource. The world cannot afford to have their efforts squandered in futile campaigns to sweep back the global tide of technology with the narrow broom of Western protest movements. The coming problems demand more subtle strategies.

No one can yet say for certain what problems will prove to be most important, or what strategies will prove best for solving them. Yet we can already see novel problems of great importance, and we can discern strategies with varying degrees of promise. In short, we can see enough about the future to identify goals worth pursuing.

PART THREE

Dangers and Hopes

11

Engines of Destruction

Nor do I doubt if the most formidable armies ever heere upon earth is a sort of soldiers who for their smallness are not visible.

—Sir WILLIAM PERRY, on microbes, 1640

REPLICATING assemblers and thinking machines pose basic threats to people and to life on Earth. Today's organisms have abilities far from the limits of the possible, and our machines are evolving faster than we are. Within a few decades they seem likely to surpass us. Unless we learn to live with them in safety, our future will likely be both exciting and short. We cannot hope to foresee all the problems ahead, yet by paying attention to the big, basic issues, we can perhaps foresee the greatest challenges and get some idea of how to deal with them.

Entire books will no doubt be written on the coming social upheavals: What will happen to the global order when assemblers and automated engineering eliminate the need for most international trade? How will society change when individuals can live indefinitely? What will we do when replicating assemblers can make almost anything without human labor? What will we do when AI systems can think faster than humans? (And before they jump to the conclusion that people will despair of doing or creating anything, the authors may consider how runners regard cars, or how painters regard cameras.)

In fact, authors have already foreseen and discussed several of these issues. Each is a matter of uncommon importance, but more fundamental than any of them is the survival of life and liberty. After all, if life or liberty is obliterated, then our ideas about social problems will no longer matter.

THE THREAT FROM THE MACHINES

In Chapter 4, I described some of what replicating assemblers will do for us if we handle them properly. Powered by fuels or sunlight, they will be able to make almost anything (including more of themselves) from common materials.

Living organisms are also powered by fuels or sunlight, and also make more of themselves from ordinary materials. But unlike assembler-based systems, they cannot make "almost anything."

Genetic evolution has limited life to a system based on DNA, RNA, and ribosomes, but memetic evolution will bring life-like machines based on nanocomputers and assemblers. I have already described how assembler-built molecular machines will differ from the ribosome-built machinery of life. Assemblers will be able to build all that ribosomes can, and more; assembler-based replicators will therefore be able to do all that life can, and more. From an evolutionary point of view, this poses an obvious threat to otters, people, cacti, and ferns—to the rich fabric of the biosphere and all that we prize.

The early transistorized computers soon beat the most advanced vacuum-tube computers because they were based on superior devices. For the same reason, early assembler-based replicators could beat the most advanced modern organisms. "Plants" with "leaves" no more efficient than today's solar cells could out-compete real plants, crowding the biosphere with an inedible foliage. Tough, omnivorous "bacteria" could out-compete real bacteria: they could spread like blowing pollen, replicate swiftly, and reduce the biosphere to dust in a matter of days. Dangerous replicators could easily be too tough, small, and rapidly spreading to stop—at least if we made no preparation. We have trouble enough controlling viruses and fruit flies.

Among the cognoscenti of nanotechnology, this threat has become known as the "gray goo problem." Though masses of uncontrolled replicators need not be gray or gooey, the term "gray goo" emphasizes that replicators able to obliterate life might be less inspiring than a single species of crabgrass. They might be "superior" in an evolutionary sense, but this need not make them valuable. We have evolved to love a world rich in living things, ideas, and diversity, so there is no reason to value gray goo merely because it could spread. Indeed, if we prevent it we will thereby prove *our* evolutionary superiority.

The gray goo threat makes one thing perfectly clear: we cannot afford certain kinds of accidents with replicating assemblers.

In Chapter 5, I described some of what advanced AI systems will do for us, if we handle them properly. Ultimately, they will embody the patterns of thought and make them flow at a pace no mammal's brain can match. AI systems that work together as people do will be able to out-think not just individuals, but whole societies. Again, the evolution of genes has left life stuck. Again, the evolution of memes by human beings—and eventually by machines—will advance our hardware far beyond the limits of life. And again, from an evolutionary point of view this poses an obvious threat.

Knowledge can bring power, and power can bring knowledge. Depending on their natures and their goals, advanced AI systems might accumulate enough knowledge and power to displace us, if we don't prepare properly. And as with replicators, mere evolutionary "superiority" need not make the victors better than the vanquished by any standard but brute competitive ability.

This threat makes one thing perfectly clear: we need to find ways to live with thinking machines, to make them law-abiding citizens.

ENGINES OF POWER

Certain kinds of replicators and AI systems may confront us with forms of hardware capable of swift, effective, independent action. But the novelty of this threat—coming from the machines themselves—must not blind us to a more traditional danger. Replicators and AI systems can also serve as great engines of power, if wielded freely by sovereign states.

Throughout history, states have developed technologies to extend their military power, and states will no doubt play a dominant role in developing replicators and AI systems. States could use replicating assemblers to build arsenals of advanced weapons, swiftly, easily, and in vast quantity. States could use special replicators directly to wage a sort of germ warfare—one made vastly more practical by programmable, computer-controlled "germs." Depending on their skills, AI systems could serve as weapon designers, strategists, or fighters. Military funds already support research in both molecular technology and artificial intelligence.

States could use assemblers or advanced AI systems to achieve sudden, destabilizing breakthroughs. I earlier discussed reasons for expecting that the advent of replicating assemblers will bring relatively sudden changes: Able to replicate swiftly, they could become abundant in a matter of days. Able to make almost anything, they could be programmed to duplicate existing weapons, but made from superior materials. Able to work with standard, well-understood components (atoms) they could suddenly build things designed in anticipation of the assembler breakthrough. These results of design-ahead could include programmable germs and other nasty novelties. For all these reasons, a state that makes the assembler breakthrough could rapidly create a decisive military force—if not literally overnight, then at least with unprecedented speed.

States could use advanced AI systems to similar ends. Automated engineering systems will facilitate design-ahead and speed assembler development. AI systems able to build better AI systems will allow an explosion of capability with effects hard to anticipate. Both AI systems and replicating assemblers will enable states to expand their military capabilities by orders of magnitude in a brief time.

Replicators can be more potent than nuclear weapons: to devastate Earth with bombs would require masses of exotic hardware and rare isotopes, but to destroy all life with replicators would require only a single speck made of ordinary elements. Replicators give nuclear war some company as a potential cause of extinction, giving a broader context to extinction as a moral concern.

Despite their potential as engines of destruction, nanotechnology and AI systems will lend themselves to more subtle uses than do nuclear weapons. A bomb can only blast things, but nanomachines and AI systems could be used to infiltrate, seize, change, and govern a territory or a world. Even the most ruthless police have no use for nuclear weapons, but they do have use for bugs, drugs, assassins, and other flexible engines of power. With advanced technology, states will be able to consolidate their power over people.

Like genes, memes, organisms, and hardware, states have evolved. Their institutions have spread (with variations) through growth, fission, imitation, and conquest. States at war fight like beasts, but using citizens as their bones, brains, and muscle. The coming breakthroughs will confront states with new pressures and opportunities, encouraging sharp changes in how states behave. This naturally gives cause for concern. States have, historically, excelled at slaughter and oppression.

In a sense, a state is simply the sum of the people making up its organizational apparatus: their actions add up to make its actions. But the same might be said of a dog and its cells, though a dog is clearly more than just a clump of cells. Both dogs and states are evolved systems, with structures that affect how their parts behave. For thousands of years, dogs have evolved largely to please people, because they have survived and reproduced at human whim. For thousands of years, states have evolved under other selective pressures. Individuals have far more power over their dogs than they do over "their" states. Though states, too, can benefit from pleasing people, their very existence has depended on their capability for *using* people, whether as leaders, police, or soldiers.

It may seem paradoxical to say that people have limited power over states: After all, aren't people behind a state's every action? But in democracies, heads of state bemoan their lack of power, representatives bow to interest groups, bureaucrats are bound by rules, and voters, allegedly in charge, curse the whole mess. The state acts and people affect it, yet no one can claim to control it. In totalitarian states, the apparatus of power has a tradition, structure, and inner logic that leaves no one free, neither the rulers nor the ruled. Even kings had to act in ways limited by the traditions of monarchy and the practicalities of power, if they were to remain kings. States are not human, though they are made of humans.

Despite this, history shows that change is possible, even change for the better. But changes always move from one semiautonomous, inhuman system to another—equally inhuman but perhaps more humane. In our hope for improvements, we must not confuse states that wear a human face with states that have humane institutions.

Describing states as quasi-organisms captures only one aspect of a complex reality, yet it suggests how they may evolve in response to the coming breakthroughs. The growth of government power, most spectacular in totalitarian countries, suggests one direction.

States could become more like organisms by dominating their parts more completely. Using replicating assemblers, states could fill the human environment with miniature surveillance devices. Using an abundance of speech-understanding AI systems, they could listen to everyone without employing half the population as listeners. Using nanotechnology like that proposed for cell repair machines, they could cheaply tranquilize, lobotomize, or otherwise modify entire populations. This would simply extend an all too familiar pattern. The world already holds governments that spy, torture, and drug; advanced technology will merely extend the possibilities.

But with advanced technology, states need not control people—they could instead simply *discard* people. Most people in most states, after all, function either as workers, larval workers, or worker-rearers, and most of these workers make, move, or grow things. A state with replicating assemblers would not need such work. What is more, advanced AI systems could replace engineers, scientists, administrators, and even leaders. The combination of nanotechnology and advanced AI will make possible intelligent, effective robots; with such robots, a state could prosper while discarding anyone, or even (in principle) everyone.

The implications of this possibility depend on whether the state exists to serve the people, or the people exist to serve the state.

In the first case, we have a state shaped by human beings to serve general human purposes; democracies tend to be at least rough approximations to this ideal. If a democratically controlled government loses its need for people, this will basically mean that it no longer needs to use people as bureaucrats or taxpayers. This will open new possibilities, some of which may prove desirable.

In the second case, we have a state evolved to exploit human beings, perhaps along totalitarian lines. States have needed people as workers because human labor has been the necessary foundation of power. What is more, genocide has been expensive and troublesome to organize and execute. Yet, in this century totalitarian states have slaughtered their citizens by the millions. Advanced technology will make workers unnecessary and genocide easy. History suggests that totalitarian states may then eliminate people wholesale. There is some consolation in this. It seems likely that a state willing and able to enslave us biologically would instead simply kill us.

The threat of advanced technology in the hands of governments makes one thing perfectly clear: we cannot afford to have an oppressive state take the lead in the coming breakthroughs.

The basic problems I have outlined are obvious: in the future, as in the past, new technologies will lend themselves to accidents and abuse. Since replicators and thinking machines will bring great new powers, the potential for accidents and abuse will likewise be great. These possibilities pose genuine threats to our lives.

Most people would like a chance to live and be free to choose how to live. This goal may not sound too utopian, at least in some parts of the world. It doesn't mean forcing everyone's life to fit some grand scheme; it chiefly means avoiding enslavement and death. Yet, like the achievement of a utopian dream, it will bring a future of wonders.

Given these life-and-death problems and this general goal, we can consider what measures might help us succeed. Our strategy must involve people, principles, and institutions, but it must also rest on tactics which inevitably will involve technology.

TRUSTWORTHY SYSTEMS

To use such powerful technologies in safety, we must make hardware we can trust. To have trust, we must be able to judge technical facts accurately, an ability that will in turn depend partly on the quality of our institutions for judgment. More fundamentally, though, it will depend on whether trustworthy hardware is physically possible. This is a matter of the reliability of components and of systems.

We can often make reliable components, even without assemblers to help. "Reliable" doesn't mean "indestructible"—anything will fail if placed close enough to a nuclear blast. It doesn't even mean "tough"—a television set may be reliable, yet not survive being bounced off a concrete floor. Rather, we call something reliable when we can count on it to work as designed.

A reliable component need not be a perfect embodiment of a perfect design: it need only be a good enough embodiment of a cautious enough design. A bridge engineer may be uncertain about the strength of winds, the weight of traffic, and the strength of steel, but by assuming high winds, heavy traffic, and weak steel, the engineer can design a bridge that will stand.

Unexpected failures in components commonly stem from material flaws. But assemblers will build components that have a negligible number of their atoms out of place—none, if need be. This will make them perfectly uniform and in a limited sense perfectly reliable. Radiation will still cause damage, though, because a cosmic ray can unexpectedly knock atoms loose from anything. In a small enough component (even in a modern computer memory device), a single particle of radiation can cause a failure.

But systems can work even when their parts fail; the key is redundancy. Imagine a bridge suspended from cables that fail randomly, each breaking about once a year at an unpredictable time. If the bridge falls when a cable breaks, it will be too dangerous to use. Imagine, though, that a broken cable takes a day to fix (because skilled repair crews with spare cables are on call), and that, though it takes five cables to support the bridge, there are actually *six*. Now if one cable breaks, the bridge will still stand. By clearing traffic and then replacing the failed cable, the bridge operators can restore safety. To destroy this bridge, a second cable must break in the same day as the first. Supported by six cables, each having a one-in-365 daily chance of breaking, the bridge will likely last about ten years.

While an improvement, this remains terrible. Yet a bridge with ten cables (five needed, five extra) will fall only if *six* cables break on the same day: the suspension system is likely to last over ten million years. With fifteen cables, the expected lifetime is over ten thousand times the age of the Earth. Redundancy can bring an exponential explosion of safety.

Redundancy works best when the redundant components are truly independent. If we don't trust the design process, then we must use components designed independently; if a bomb, bullet, or cosmic ray may damage several neighboring parts, then we must spread redundant parts more widely. Engineers who want to supply reliable transportation between two islands shouldn't just add more cables to a bridge. They should build two well-separated bridges using different designs, then add a tunnel, a ferry, and a pair of inland airports.

Computer engineers also use redundancy. Stratus Computer Inc., for example, makes a machine that uses four central processing units (in two pairs) to do the work of one, but to do it vastly more reliably.

Each pair is continually checked for internal consistency, and a failed pair can be replaced while its twin carries on.

An even more powerful form of redundancy is *design diversity*. In computer hardware, this means using several computers with *different designs*, all working in parallel. Now redundancy can correct not just for failures in a piece of hardware, but for errors in its design.

Much has been made of the problem of writing large, error-free programs; many people consider such programs impossible to develop and debug. But researchers at the UCLA Computer Science Department have shown that design diversity can also be used in software: several programmers can tackle the same problem independently, then all their programs can be run in parallel and made to vote on the answer. This multiplies the cost of writing and running the program, but it makes the resulting software system resistant to the bugs that appear in some of its parts.

We can use redundancy to control replicators. Just as repair machines that compare multiple DNA strands will be able to correct mutations in a cell's genes, so replicators that compare multiple copies of their instructions (or that use other effective error-correcting systems) will be able to resist mutation in these "genes." Redundancy can again bring an exponential explosion of safety.

We can build systems that are extremely reliable, but this will entail costs. Redundancy makes systems heavier, bulkier, more expensive, and less efficient. Nanotechnology, though, will make most things far lighter, smaller, cheaper, and more efficient to begin with. This will make redundancy and reliability more practical.

Today, we are seldom willing to pay for the safest possible systems; we tolerate failures more-or-less willingly and seldom consider the real limits of reliability. This biases judgments of what can be achieved. A psychological factor also distorts our sense of how reliable things can be made: failures stick in our minds, but everyday successes draw little attention. The media amplify this tendency by reporting the most dramatic failures from around the world, while ignoring the endless and boring successes. Worse yet, the components of redundant systems may fail in visible ways, stirring alarums: imagine how the media would report a snapped bridge cable, even if the bridge were the super-safe fifteen-cable model described above. And since each added redundant component adds to the chance of a component failure, a system's reliability can *seem* worse even as it approaches perfection.

Appearances aside, redundant systems made of abundant, flawless components can often be made almost perfectly reliable. Redundant systems spread over wide enough spaces will survive even bullets and bombs.

But what about design errors? Having a dozen redundant parts will do no good if they share a fatal error in design. Design diversity is one anwer; good testing is another. We can reliably evolve good designs without being reliably good designers: we need only be good at testing, good at tinkering, and good at being patient. Nature has evolved working molecular machinery through entirely mindless tinkering and testing. Having minds, we can do as well or better.

We will find it easy to design reliable hardware if we can develop reliable automated engineering systems. But this raises the wider issue of developing trustworthy artificial intelligence systems. We will have little trouble making AI systems with reliable hardware, but what about their software?

Like present AI systems and human minds, advanced AI systems will be synergistic combinations of many simpler parts. Each part will be more specialized and less intelligent than the system as a whole. Some parts will look for patterns in pictures, sounds, and other data and suggest what they might mean. Other parts will compare and judge the suggestions of these parts. Just as the pattern recognizers in the human visual system suffer from errors and optical illusions, so will the pattern recognizers in AI systems. (Indeed, some advanced machine vision systems already suffer from familiar optical illusions.) And just as other parts of the human mind can often identify and compensate for illusions, so will other parts of AI systems.

As in human minds, intelligence will involve mental parts that make shaky guesses and other parts that discard most of the bad guesses before they draw much attention or affect important decisions. Mental parts that reject action ideas on ethical grounds correspond to what we call a conscience. AI systems with many parts will have room for redundancy and design diversity, making reliability possible.

A genuine, flexible AI system must evolve ideas. To do this, it must find or form hypotheses, generate variations, test them, and then modify or discard those found inadequate. Eliminating any of these abilities would make it stupid, stubborn, or insane ("Durn machine can't think and won't learn from its mistakes—junk it!"). To avoid becoming trapped by initial misconceptions, it will have to consider conflicting views, seeing how well each explains the data, and seeing whether one view can explain another.

Scientific communities go through a similar process. And in a paper called "The Scientific Community Metaphor," William A. Kornfeld and Carl Hewitt of the MIT Artificial Intelligence Laboratory suggest that AI researchers model their programs still more closely on the evolved structure of the scientific community. They point to the pluralism of science, to its diversity of competing proposers, supporters, and critics. Without proposers, ideas cannot appear; without supporters, they cannot grow; and without critics to weed them, bad ideas can crowd out the good. This holds true in science, in technology, in AI systems, and among the parts of our own minds.

Having a world full of diverse and redundant proposers, supporters, and critics is what makes the advance of science and technology reliable. Having more proposers makes good proposals more common; having more critics makes bad proposals more vulnerable. Better, more numerous ideas are the result. A similar form of redundancy can help AI systems to develop sound ideas.

People sometimes guide their actions by standards of truth and ethics, and we should be able to evolve AI systems that do likewise, but more reliably. Able to think a million times faster than us, they will have more time for second thoughts. It seems that AI systems can be made trustworthy, at least by human standards.

I have often compared AI systems to individual human minds, but the resemblance need not be close. A system that can mimic a person may need to be personlike, but an automated engineering system probably doesn't. One proposal (called an Agora system, after the Greek term for a meeting and market place) would consist of many independent pieces of software that interact by offering one another services in exchange for money. Most pieces would be simple-minded specialists, some able to suggest a design change, and others able to analyze one. Much as Earth's ecology has evolved extraordinary organisms, so this computer economy could evolve extraordinary designs—and perhaps in a comparably mindless fashion. What is more, since the system would be spread over many machines and have parts written by many people, it could be diverse, robust, and hard for any group to seize and abuse.

Eventually, one way or another, automated engineering systems will be able to design things more reliably than any group of human engineers can today. Our challenge will be to design *them* correctly.

We will need human institutions that reliably develop reliable systems.

Human institutions are evolved artificial systems, and they can often solve problems that their individual members cannot. This makes them a sort of "artificial intelligence system." Corporations, armies, and research laboratories all are examples, as are the looser structures of the market and the scientific community. Even governments may be seen as artificial intelligence systems—gross, sluggish, and befuddled, yet superhuman in their sheer capability. And what are constitutional checks and balances but an attempt to increase a government's reliability through institutional diversity and redundancy? When we build intelligent machines, we will use them to check and balance one another.

By applying the same principles, we may be able to develop reliable, technically oriented institutions having strong checks and balances, then use these to guide the development of the systems we will need to handle the coming breakthroughs.

TACTICS FOR THE ASSEMBLER BREAKTHROUGH

Some force in the world (whether trustworthy or not) will take the lead in developing assemblers; call it the "leading force." Because of the strategic importance of assemblers, the leading force will presumably be some organization or institution that is effectively controlled by some government or group of governments. To simplify matters, pretend for the moment that we (the good guys, attempting to be wise) can make policy for the leading force. For citizens of democratic states, this seems a good attitude to take.

What should we do to improve our chances of reaching a future worth living in? What *can* we do?

We can begin with what must *not* happen: we must not let a single replicating assembler of the wrong kind be loosed on an unprepared world. Effective preparations seem possible (as I will describe), but it seems that they must be based on assembler-built systems that can be built only after dangerous replicators have already become possible. Design-ahead can help the leading force prepare, yet even vigorous, foresighted action seems inadequate to prevent a time of danger. The reason is straightforward: dangerous replicators will be far simpler to design than systems that can thwart them, just as a bacterium is far simpler than an immune system. We will need tactics for containing nanotechnology while we learn how to tame it.

One obvious tactic is isolation: the leading force will be able to contain replicator systems behind multiple walls or in laboratories in space. Simple replicators will have no intelligence, and they won't be designed to escape and run wild. Containing them seems no great challenge.

Better yet, we will be able to design replicators that *can't* escape and run wild. We can build them with counters (like those in cells) that limit them to a fixed number of replications. We can build them to have requirements for special synthetic "vitamins," or for bizarre environments found only in the laboratory. Though replicators could be made tougher and more voracious than any modern pests, we can also make them useful but harmless. Because we will design them from scratch, replicators need not have the elementary survival skills that evolution has built into living cells.

Further, they need not be able to evolve. We can give replicators redundant copies of their "genetic" instructions, along with repair mechanisms to correct any mutations. We can design them to stop working long before enough damage accumulates to make a lasting mutation a significant possibility. Finally, we can design them in ways that would hamper evolution even if mutations *could* occur.

Experiments show that most computer programs (other than specially designed AI programs, such as Dr. Lenat's EURISKO) seldom respond to mutations by changing slightly; instead, they simply fail. Because they cannot vary in useful ways, they cannot evolve. Unless they are specially designed, replicators directed by nanocomputers will share this handicap. Modern organisms are fairly good at evolving partly because they descend from ancestors that evolved. They are *evolved* to evolve; this is one reason for the complexities of sexual reproduction and the shuffling of chromosome segments during the production of sperm and egg cells. We can simply neglect to give replicators similar talents.

It will be easy for the leading force to make replicating assemblers useful, harmless, and stable. Keeping assemblers from being stolen and abused is a different and greater problem, because it will be a game played against intelligent opponents. As one tactic, we can reduce the incentive to steal assemblers by making them available in safe forms. This will also reduce the incentive for other groups to develop assemblers independently. The leading force, after all, will be followed by trailing forces.

Limited Assemblers

In Chapter 4, I described how a system of assemblers in a vat could build an excellent rocket engine. I also pointed out that we will be able to make assembler systems that act like seeds, absorbing sunlight and ordinary materials and growing to become almost anything. These special-purpose systems will not replicate themselves, or will do so only a fixed number of times. They will make only *what* they were programmed to make, *when* they are told to make it. Anyone lacking special assembler-built tools would be unable to reprogram them to serve other purposes.

Using limited assemblers of this sort, people will be able to make as much as they want of whatever they want, subject to limits built into the machines. If none is programmed to make nuclear weapons, none will; if none is programmed to make dangerous replicators, none will. If some are programmed to make houses, cars, computers, toothbrushes, and whatnot, then these products can become cheap and abundant. Machines built by limited assemblers will enable us to open space, heal the biosphere, and repair human cells. Limited assemblers can bring almost unlimited wealth to the people of the world.

This tactic will ease the moral pressure to make unlimited assemblers available immediately. But limited assemblers will still leave legitimate needs unfulfilled. Scientists will need freely programmable assemblers to conduct studies; engineers will need them to test designs. These needs can be served by sealed assembler laboratories.

Sealed Assembler Laboratories

Picture a computer accessory the size of your thumb, with a state-of-the-art plug on its bottom. Its surface looks like boring gray plastic, imprinted with a serial number, yet this sealed assembler lab is an assembler-built object that contains many things. Inside, sitting above the plug, is a large nanoelectronic computer running advanced molecular-simulation software (based on the software developed during assembler development). With the assembler lab plugged in and turned on, your assembler-built home computer displays a three-dimensional picture of whatever the lab computer is simulating, representing atoms as colored spheres. With a joystick, you can direct the simulated assembler arm to build things. Programs can move the arm faster, building elaborate structures on the screen in the blink of an eye. The simulation always works perfectly, because the nanocomputer cheats: as you make the simulated arm move simulated molecules, the computer directs an actual arm to move actual molecules. It then checks the results whenever needed to correct its calculations.

The end of this thumb-sized object holds a sphere built in many concentric layers. Fine wires carry power and signals through the layers; these let the nanocomputer in the base communicate with the devices at the sphere's center. The outermost layer consists of sensors. Any attempt to remove or puncture it triggers a signal to a layer near the core. The next layer in is a thick spherical shell of prestressed diamond composite, with its outer layers stretched and its inner layers compressed. This surrounds a layer of thermal insulator which in turn surrounds a peppercorn-sized spherical shell made up of microscopic, carefully arranged blocks of metal and oxidizer. These are laced with electrical igniters. The outer sensor layer, if punctured, triggers these igniters. The metal-and-oxidizer demolition charge then burns in a fraction of a second, producing a gas of metal oxides denser than water and almost as hot as the surface of the Sun. But the blaze is tiny; it swiftly cools, and the diamond sphere confines its great pressure.

This demolition charge surrounds a smaller composite shell, which surrounds another layer of sensors, which can also trigger the demolition charge. These sensors surround the cavity which contains the actual sealed assembler lab.

These elaborate precautions justify the term "sealed." Someone outside cannot open the lab space without destroying the contents, and *no assemblers or assembler-built structures can escape from within.* The system is designed to let out information, but not dangerous replicators or dangerous tools. Each sensor layer consists of many redundant layers of sensors, each *intended* to detect any possible penetration, and each making up for possible flaws in the others. Penetration, by triggering the demolition charge, raises the lab to a temperature beyond the melting point of all possible substances and makes the survival of a dangerous device impossible. These protective mechanisms all gang up on something about a millionth their size—that is, on whatever will fit in the lab, which provides a spherical work space no wider than a human hair.

Though small by ordinary standards, this work space holds room enough for millions of assemblers and thousands of trillions of atoms. These sealed labs will let people build and test devices, even voracious replicators, in complete safety. Children will use the atoms inside them as a construction set with almost unlimited parts. Hobbyists will exchange programs for building various gadgets. Engineers will build and test new nanotechnologies. Chemists, materials scientists, and biologists will build apparatus and run experiments. In labs built around biological samples, biomedical engineers will develop and test early cell repair machines.

In the course of this work, people will naturally develop useful designs, whether for computer circuits, strong materials, medical devices, or whatever. After a public review of their safety, these things could be made available outside the sealed labs by programming limited assemblers to make them. Sealed labs and limited assemblers will form a complementary pair: The first will let us invent freely; the second will let us enjoy the fruits of our invention safely. The chance to pause between design and release will help us avoid deadly surprises.

Sealed assembler labs will enable the whole of society to apply its creativity to the problems of nanotechnology. And this will speed our preparations for the time when an independent force learns how to build something nasty.

Hiding Information

In another tactic for buying time, the leading force can *attempt* to burn the bridge it built from bulk to molecular technology. This means destroying the records of how the first assemblers were made (or making the records thoroughly inaccessible). The leading force may be able to develop the first, crude assemblers in such a way that no one knows the details of more than a small fraction of the whole system. Imagine that we develop assemblers by the route outlined in Chapter 1. The protein machines that we use to build the first crude assemblers will then promptly become obsolete. If we destroy the records of the protein designs, this will hamper efforts to duplicate them, yet will not hamper further progress in nanotechnology.

If sealed labs and limited assemblers are widely available, people will have little scientific or economic motivation to redevelop nanotechnology independently, and burning the bridge from bulk technology will make independent development more difficult. Yet these can be no more than delaying tactics. They won't stop independent development; the human urge for power will spur efforts which will eventually succeed. Only detailed, universal policing on a totalitarian scale could stop independent development indefinitely. If the policing were conducted by anything like a modern government, this would be a cure roughly as dangerous as the disease. And even then, would people maintain perfect vigilance *forever?*

It seems that we must eventually learn to live in a world with untrustworthy replicators. One sort of tactic would be to hide behind a wall or to run far away. But these are brittle methods: dangerous replicators might breach the wall or cross the distance, and bring utter disaster. And, though walls can be made proof against small replicators, no fixed wall will be proof against large-scale, organized malice. We will need a more robust, flexible approach.

Active Shields

It seems that we can build nanomachines that act somewhat like the white blood cells of the human immune system: devices that can fight not just bacteria and viruses, but dangerous replicators of all sorts. Call an automated defense of this sort an *active shield,* to distinguish it from a fixed wall.

Unlike ordinary engineering systems, reliable active shields must do more than just cope with nature or clumsy users. They must also cope with a far greater challenge—with the entire range of threats that intelligent forces can design and build under prevailing circumstances. Building and improving prototype shields will be akin to running both sides of an arms race on a laboratory scale. But the goal here will be to seek the minimum requirements for a defense that reliably prevails.

In Chapter 5, I described how Dr. Lenat and his EURISKO program evolved successful fleets to fight according to the rules of a naval-warfare simulation game. In a similar way, we can make into a game the deadly serious effort to develop reliable shields, using sealed assembler labs of various sizes as playing fields. We can turn loose a horde of engineers, computer hackers, biologists, hobbyists, and automated engineering systems, all invited to pit their systems against one another in games limited only by the initial conditions, the laws of nature, and the walls of the sealed labs. These competitors will evolve threats and shields in an open-ended series of microbattles. When replicating assemblers have brought abundance, people will have time enough for so important a game.

Eventually we can test promising shield systems in Earthlike environments in space. Success will make possible a system able to protect human life and Earth's biosphere from the worst that a fistful of loose replicators can do.

IS SUCCESS POSSIBLE?

With our present uncertainties, we cannot yet describe either threats or shields with any accuracy. Does this mean we can't have any confidence that effective shields are possible? Apparently we can; there is a difference, after all, between knowing that something is possible and knowing how to do it. And in this case, the world holds examples of analogous successes.

There is nothing fundamentally novel about defending against invading replicators; life has been doing it for ages. Replicating assemblers, though unusually potent, will be physical systems not unlike those we already know. Experience suggests that they can be controlled.

Viruses are molecular machines that invade cells; cells use molecular machines (such as restriction enzymes and antibodies) to defend against them. Bacteria are cells that invade organisms; organisms use cells (such as white blood cells) to defend against them. Similarly, societies use police to defend against criminals and armies to defend against invaders. On a less physical level, minds use meme systems such as the scientific method to defend against nonsense, and societies use institutions such as courts to defend against the power of other institutions.

The biological examples in the last paragraph show that even after a billion-year arms race, molecular machines have maintained successful defenses against molecular replicators. Failures have been common too, but the successes do indicate that defense is possible. These successes suggest that we can indeed use nanomachines to defend against nanomachines. Though assemblers will bring many advances, there seems no reason why they should permanently tip the balance against defense.

The examples given above—some involving viruses, some involving institutions—are diverse enough to suggest that successful defense rests on general principles. One might ask, Why do all these defenses succeed? But turn the question around: Why should they fail? Each conflict pits similar systems against each other, giving the attacker no obvious advantage. In each conflict, moreover, the attacker faces a defense *that is well established*. The defenders fight on home ground, giving them advantages such as prepared positions, detailed local knowledge, stockpiled resources, and abundant allies —when the immune system recognizes a germ, it can mobilize the resources of an entire body. All these advantages are general and basic, having little to do with the details of a technology. We can give our active shields the same advantages over dangerous replicators. And they need not sit idle while dangerous weapons are amassed, any more than the immune system sits idle while bacteria multiply.

It would be hard to predict the outcome of an open-ended arms race between powers equipped with replicating assemblers. But before this situation can arise, the leading force seems likely to acquire a temporary but overwhelming military advantage. If the outcome of an arms race is in doubt, then the leading force will likely use its strength to ensure that no opponents are allowed to catch up. If it does so, then active shields will not have to withstand attacks backed by the resources of half a continent or half a solar system; they will instead be like a police force or an immune system, facing attacks backed only by whatever resources can be gathered in secret within the protected territory.

In each case of successful defense that I cited above, the attackers and the shields have developed through broadly similar processes. The immune system, shaped by genetic evolution, meets threats also shaped by genetic evolution. Armies, shaped by human minds, also meet similar threats. Likewise, both active shields and dangerous replicators will be shaped by memetic evolution. But if the leading force can develop automated engineering systems that work a millionfold faster than human engineers, and if it can use them for a single year, then it can build active shields based on a million years' worth of engineering advance. With such systems we may be able to explore the limits of the possible well enough to build a reliable shield against all physically possible threats.

Even without our knowing the details of the threats and the shields, there seems reason to believe that shields are possible. And the examples of memes controlling memes and of institutions controlling institutions also suggest that AI systems can control AI systems.

In building active shields, we will be able to use the power of replicators and AI systems to multiply the traditional advantages of the defending force: we can give it overwhelming strength through abundant, replicator-built hardware with designs based on the equivalent of a million-year lead in technology. We can build active shields having strength and reliability that will put past systems to shame.

Nanotechnology and artificial intelligence could bring the ultimate tools of destruction, but they are not inherently destructive. With care, we can use them to build the ultimate tools of peace.

12

Strategies and Survival

He that will not apply new remedies must expect new evils; for time is the greatest innovator.

—FRANCIS BACON

IN EARLIER CHAPTERS I have stuck close to the firm ground of technological possibility. Here, however, I must venture further into the realm of politics and human action. This ground is softer, but technological facts and evolutionary principles still provide firm points on which to stand and survey the territory.

The technology race, driven by evolutionary pressures, is carrying us toward unprecedented dangers; we need to find strategies for dealing with them. Since we see such great dangers ahead, it makes sense to consider stopping our headlong rush. But how can we?

Personal Restraint

As individuals, we could refrain from doing research that leads toward dangerous capabilities. Indeed, most people *will* refrain, since most are not researchers in the first place. But this strategy won't stop advances: in our diverse world, others will carry the work forward.

Local Suppression

A strategy of personal restraint (at least in this matter) smacks of simple inaction. But what about a strategy of local political action, of lobbying for laws to suppress certain kinds of research? This would be personal action aimed at enforcing *collective* inaction. Although it might succeed in suppressing research in a city, a district, a country, or an alliance, this strategy cannot help us guide the leading force; instead, it would let some force beyond our control take the lead. A popular movement of this sort can halt research only where the people hold the power, and its greatest possible success would merely open the way for a more repressive state to become the leading force.

Where nuclear weapons are concerned, arguments can be made for unilateral disarmament and nonviolent (or at least non-nuclear) resistance. Nuclear weapons can be used to smash military establishments and spread terror, but they cannot be used to occupy territory or rule people—not directly. Nuclear weapons have failed to suppress guerrilla warfare and social unrest, so a strategy of disarmament and resistance makes some degree of sense.

The unilateral suppression of nanotechnology and AI, in contrast, would amount to unilateral disarmament in a situation where resistance cannot work. An aggressive state could use these technologies to seize and rule (or exterminate) even a nation of Gandhis, or of armed and dedicated freedom fighters.

This deserves emphasis. Without some novel way to reform the world's oppressive states, simple research-suppression movements cannot have total success. Without a *total* success, a *major* success would mean disaster for the democracies. Even if they got nowhere, efforts of this sort would absorb the work and passion of activists, wasting scarce human resources on a futile strategy. Further, efforts at suppression would alienate concerned researchers, stirring fights between potential allies and wasting further human resources. Its futility and divisiveness make this a strategy to be shunned.

Nonetheless, suppression has undeniable appeal. It is simple and direct: "Danger coming? Let's stop it!" Further, successes in local lobbying efforts promise short-term gratification: "Danger coming? We can stop it *here* and *now,* for a start!" The start would be a false start, but not everyone will notice. The idea of simple suppression seems likely to seduce many minds. After all, local suppression of local dangers has a long, successful tradition; stopping a local polluter, for example, reduces local pollution. Efforts at local suppression of global dangers will *seem* similar, however different the effects may be. We will need local organization and political pressure, but they must be built around a workable strategy.

Global Suppression Agreements

In a more promising approach, we could apply local pressure for the negotiation of a verifiable, worldwide ban. A similar strategy might have a chance in the control of nuclear weapons. But stopping nanotechnology and artificial intelligence would pose problems of a different order, for at least two reasons.

First, these technologies are less well-defined than nuclear weapons: because current nuclear technology demands certain isotopes of rare metals, it is distinct from other activities. It can be defined and (in principle) banned. But modern biochemistry leads in small steps to nanotechnology, and modern computer technology leads in small steps to AI. No line defines a natural stopping point. And since each small advance will bring medical, military, and economic benefits, how could we negotiate a worldwide agreement on where to stop?

Second, these technologies are more potent than nuclear weapons: because reactors and weapons systems are fairly large, inspection could limit the size of a secret force and thus limit its strength. But dangerous replicators will be microscopic, and AI software will be intangible. How could anyone be sure that some laboratory somewhere isn't on the verge of a strategic breakthrough? In the long run, how could anyone even be sure that some hacker in a basement isn't on the verge of a strategic breakthrough? Ordinary verification measures won't work, and this makes negotiation and enforcement of a worldwide ban almost impossible.

Pressure for the right kinds of international agreements will make our path safer, but agreements simply to suppress dangerous advances apparently won't work. Again, local pressure must be part of a *workable* strategy.

Global Suppression by Force

If peaceful agreements won't work, one might consider using military force to suppress dangerous advances. But because of verification problems, military pressure alone would not be enough. To suppress advances by force would instead require that one power conquer and occupy hostile powers armed with nuclear weapons—hardly a safe policy. Further, the conquering power would itself be a major technological force with massive military power and a demonstrated willingness to use it. Could this power then be trusted to suppress *its own* advances? And even if so, could it be trusted to maintain unending, omnipresent vigilance over the whole world? If not, then threats will eventually emerge in secret, and in a world where open work on active shields has been prevented. The likely result would be disaster.

Military strength in the democracies has great benefits, but military strength alone cannot solve our problem. We cannot win safety through a strategy of conquest and research suppression.

These strategies for stopping research—whether through personal inaction, local inaction, negotiated agreement, or world conquest—all seemed doomed to fail. Yet opposition to advances will have a role to play, because we will need selective, intelligently targeted delay to postpone threats until we are prepared for them. Pressure from alert activists will be essential, but to help guide advance, not to halt it.

Unilateral Advance

If attempts to suppress research in AI and nanotechnology seem futile and dangerous, what of the opposite course—an all-out, unilateral effort? But this too presents problems. We in the democracies probably cannot produce a major strategic breakthrough in perfect secrecy. Too many people would be involved for too many years. Since the Soviet leadership would learn of our efforts, their reaction becomes an obvious concern, and they would surely view a great breakthrough on our part as a great threat. If nanotechnology were being developed as part of a secret military program, their intelligence analysts would fear the development of a subtle but decisive weapon, perhaps based on programmable "germs." Depending on the circumstances, our opponents might choose to attack while they still could. It is important that the democracies keep the lead in these technologies, but we will be safest if we can somehow combine this strength with clearly nonthreatening policies.

Balance of Power

If we follow any of the strategies above we will inevitably stir strong conflict. Attempts to suppress nanotechnology and AI will pit the would-be suppressors against the vital interests of researchers, corporations, military establishments, and medical patients. Attempts to gain unilateral advantage through these technologies will pit the cooperating democracies against the vital interests of our opponents. All strategies will stir conflict, but need all strategies split Western societies or the world so badly?

In search of a middle path, we might seek a balance of power based on a balance of technology. This would seemingly extend a situation that has kept a measure of peace for four decades. But the key word here is "seemingly": the coming breakthroughs will be too abrupt and destabilizing for the old balance to continue. In the past, a country could suffer a technological lag of several years and yet maintain a rough military balance. With swift replicators or advanced AI, though, a lag of a single *day* could be fatal. A stable balance seems too much to hope for.

Cooperative Development

There is, in principle, a way to ensure a technological balance between the cooperating democracies and Soviet bloc: we could develop the technologies cooperatively, sharing our tools and information. Though this has obvious problems, it is at least *somewhat* more practical than it may sound.

Is cooperation possible to negotiate? Failed attempts to negotiate effective arms control treaties immediately leap to mind, and cooperation may seem even more complicated and difficult to arrange. But is it? In arms control, each side is attempting to hinder the other's actions; this reinforces their adversarial relationship. Further, it stirs conflicts within each camp between groups that favor arms limitation and groups that exist to build arms. Worse yet, the negotiations revolve around words and their meanings, but each side has its own language and an incentive to twist meanings to suit itself.

Cooperation, in contrast, involves both sides working toward a shared goal; this tends to blur the adversarial nature of the relationship. Further, it may lessen the conflicts within each camp, since cooperative efforts would create projects, not destroy them. Finally, both sides discuss their efforts in a shared language—the language of mathematics and diagrams used in science and engineering. Also, cooperation has clear-cut, visible results. In the mid-1970s, the U.S. and U.S.S.R. flew a joint space mission, and until political tensions grew they were laying tentative plans for a joint space station. These were not isolated incidents, in space or on the ground; joint projects and technical exchange have gone on for years. For all its problems, technological cooperation has proved at least as easy as arms control —perhaps even easier, considering the great effort poured into the latter.

Curiously, where AI and nanotechnology are concerned, cooperation and effective arms control would have a basic similarity. To verify an arms control agreement would require constant, intimate inspection of each side's laboratories by the other's experts—a relationship as close as the most thorough cooperation imaginable.

But what would cooperation accomplish? It might ensure balance, but balance will not ensure stability. If two gunmen face each other with weapons drawn and fears high, their power is balanced, but the one that shoots first can eliminate the other's threat. A cooperative effort in technology development, unless carefully planned and controlled, would give each side fearsome weapons while providing neither side with a shield. Who could be sure that neither side would find a way to strike a disarming blow with impunity?

And even if one could guarantee this, what about the problem of other powers—and hobbyists, and accidents?

In the last chapter I described a solution to these problems: the development, testing, and construction of active shields. They offer a new remedy for a new problem, and no one has yet suggested a plausible alternative to them. Until someone does, it seems wise to consider how they might be built and whether they might make possible a strategy that can work.

A SYNTHESIS OF STRATEGIES

Personal restraint, local action, selective delay, international agreement, unilateral strength, and international cooperation—all these strategies can help us in an effort to develop active shields.

Consider our situation today. The democracies have for decades led the world in most areas of science and technology; we lead today in computer software and biotechnology. Together, we *are* the leading force. There seems no reason why we cannot maintain that lead and use it.

As discussed in the last chapter, the leading force will be able to use several tactics to handle the assembler breakthrough. These include using sealed assembler labs and limited assemblers, and maintaining secrecy regarding the details of initial assembler development. In the time we buy through these (and other) policies, we can work to develop active shields able to give us permanent protection against the new dangers. This defines a goal. To reach it, a two-part strategy seems best.

The first part involves action within the cooperating democracies. We need to maintain a lead that is comfortable enough for us to proceed with caution; if we felt we might lose the race, we might well panic. Proceeding with caution means developing trustworthy institutions for managing both the initial breakthroughs and the development of active shields. The shields we develop, in turn, must be designed to help us secure a future worth living in, a future with room for diversity.

The second part of this strategy involves policies toward presently hostile powers. Here, our aim must be to keep the initiative while minimizing the threat we present. Technological balance will not work, and we cannot afford to give up our lead. This leaves strength and leadership as our only real choice, making a nonthreatening posture doubly difficult to achieve. Here again we have need for stable, trustworthy institutions: if we can give them a great built-in inertia regarding their objectives, then perhaps even our opponents will have a measure of confidence in them.

To reassure our opponents (and ourselves!) these institutions should be as open as possible, consistent with their mission. We may also manage to build institutions that offer a role for Soviet cooperation. By inviting their participation, even if they refuse the terms we offer, we would offer some measure of reassurance regarding our intentions. If the Soviets were to accept, they would gain a public stake in our joint success.

Still, if the democracies are strong when the breakthroughs approach, and if we avoid threatening any government's control over its own territory, then our opponents will presumably see no advantage in attacking. Thus, we can probably do without cooperation, if necessary.

ACTIVE SHIELDS VS. SPACE WEAPONS

It may be useful to consider how we might apply the idea of active shields in more conventional fields. Traditionally, defense has required weapons that are also useful for offense. This is one reason why "defense" has come to mean "war-making ability," and why "defense" efforts give opponents reason for fear. Presently proposed space-based defenses will extend this pattern. Almost any defensive system that can destroy attacking missiles could also destroy an opponent's defenses—or enforce a space blockade, preventing an opponent from building defenses in the first place. Such "defenses" smell of offense, as seemingly they must, to do their job. And so the arms race gathers itself for another dangerous leap.

Must defense and offense be so nearly inseparable? History makes it seem so. Walls only halt invaders when defended by warriors, but warriors can themselves march off to invade other lands. When we picture a weapon, we naturally picture human hands aiming it and human whim deciding when to fire—and history has taught us to fear the worst.

Yet today, for the first time in history, we have learned how to build defensive systems that are fundamentally different from such weapons. Consider a space-based example. We now can design devices that sense *(looks like a thousand missiles have just been launched)*, assess *(this looks like an attempted first strike)* and act *(try to destroy those missiles!)*. If a system will fire *only* at massive flights of missiles, then it *cannot* be used for offense or a space blockade. Better yet, it could be made incapable of discriminating between attacking sides. Though serving the strategic interests of its builders, it would not be subject to the day-to-day command of anyone's generals. It would just make space a hazardous environment for an attacker's missiles. Like a sea or a mountain range in earlier wars, it would threaten neither side while providing each with some protection against the other.

Though it would use weapons technologies (sensors, trackers, lasers, homing projectiles, and such), this defense wouldn't be a weapons system, because its role would be fundamentally different. Systems of this sort need a distinctive name: they are, in fact, a sort of active shield—a term that can describe any automated or semiautomated system designed to protect without threatening. By defending both sides while threatening neither, active shields could weaken the cycle of the arms race.

The technical, economic, and strategic issues raised by active shields are complex, and they may or may not be practical in the preassembler era. If they *are* practical, then there will be several possible approaches to building them. In one approach, the cooperating democracies would build shields unilaterally. To enable other nations to verify what the system will and (more important) *won't* do, we could allow multilateral inspection of key designs, components, and production steps. We needn't give away all the technologies involved, because know-what isn't the same as know-how. In a different approach, we would build shields jointly, limiting technology transfer to the minimum required for cooperation and verification (using principles discussed in the Notes).

We have more chance of banning space weapons than we do of banning nanotechnology, and this might even be the best way to minimize our near-term risks. In choosing a long-term strategy for controlling the arms race, though, we must consider more than the next step. The analysis I have outlined in this chapter suggests that traditional arms control approaches, based on negotiating verifiable limitations, cannot cope with nanotechnology. If this is the case, then we need to develop alternative approaches. Active shields—which seem essential, eventually—may offer a new, stabilizing alternative to an arms race in space. By exploring this alternative, we can explore basic issues common to all active shields. If we then develop them, we will gain experience and build institutional arrangements that may later prove essential to our survival.

Active shields are a new option based on new technologies. Making them work will require a creative, interdisciplinary synthesis of ideas in engineering, strategy, and international affairs. They offer fresh choices that may enable us to avoid old impasses. They apparently offer an answer to the ancient problem of protecting without threatening—but not an *easy* answer.

POWER, EVIL, INCOMPETENCE, AND SLOTH

I have outlined how nanotechnology and advanced AI will give great power to the leading force—power that can be used to destroy life, or to extend and liberate it. Since we cannot stop these technologies, it seems that we must somehow cope with the emergence of a concentration of power greater than any in history.

We will need a suitable system of institutions. To handle complex technologies safely, this system must have ways to judge the relevant facts. To handle great power safely, it must incorporate effective checks and balances, and its purposes and methods must be kept open to public scrutiny. Finally, since it will help us lay the foundations for a new world, it had best be guided by our shared interests, within a framework of sound principles.

We won't start from scratch; we will build on the institutions we have. They are diverse. Not all of our institutions are bureaucracies housed in massive gray buildings; they include such diffuse and lively institutions as the free press, the research community, and activist networks. These decentralized institutions help us control the gray, bureaucratic machines.

In part, we face a new version of the ancient and general problem of limiting the abuse of power. This presents no great, fundamental novelty, and the centuries-old principles and institutions of liberal democracy suggest how it may be solved. Democratic governments already have the physical power to blast continents and to seize, imprison, and kill their citizens. But we can live with these capabilities because these governments are fairly tame and stable.

The coming years will place greater burdens on our institutions. The principles of representative government, free speech, due process, the rule of law, and protection of human rights will remain crucial. To prepare for the new burdens, we will need to extend and reinvigorate these principles and the institutions that support them; protecting free speech regarding technical matters may be crucial. Though we face a great challenge, there is reason to hope that we can meet it.

There are also, of course, obvious reasons for doubting that we can meet it. But despair is contagious and obnoxious and leaves people depressed. Besides, despair seems unjustified, despite familiar problems: *Evil*—are we too wicked to do the right thing? *Incompetence*—are we too stupid to do the right thing? *Sloth*—are we too lazy to prepare?

While it would be rash to predict a rosy future, these problems do not seem insurmountable.

Democratic governments are big, sloppy, and sometimes responsible for atrocities, yet they do not seem *evil,* as a whole, though they may contain people who deserve the label. In fact, their leaders gain power largely by appearing to uphold conventional ideas of good. Our chief danger is that policies that *seem* good may lead to disaster, or that truly good policies won't be found, publicized, and implemented in time to take effect. Democracies suffer more from sloth and incompetence than from evil.

Incompetence will of course be inevitable, but need it be fatal? We human beings are by nature stupid and ignorant, yet we sometimes manage to combine our scraps of competence and knowledge to achieve great things. No one knew how to get to the Moon, and no *one* ever learned, yet a dozen people have walked its surface. We have succeeded in technical matters because we have learned to build institutions that draw many people together to generate and test ideas. These institutions gain reliability through redundancy, and the quality of their results depends largely on how much we care and how hard we work. When we focus enough attention and resources on reliability, we often succeed. This is why the Moon missions succeeded without a fatality in space, and why no nuclear weapon has ever been launched or detonated by accident. And this is why we *may* manage to handle nanotechnology and advanced AI with sufficient care to ensure a competent job. Erratic people of limited competence can join to form stable, competent institutions.

Sloth—intellectual, moral, and physical—seems perhaps our greatest danger. We can only meet great challenges with great effort. Will enough people make enough effort? No one can say, because no one can speak for everyone else. But success will not require a sudden, universal enlightenment and mobilization. It will require only that a growing community of people strive to develop, publicize, and implement workable solutions—and that they have a good and growing measure of success.

This is not so implausible. Concern about technology has become widespread, as has the idea that accelerating change will demand better foresight. Sloth will not snare everyone, and misguided thinkers will not misdirect everyone's effort. Deadly pseudo-solutions (such as blocking research) will lose the battle of ideas if enough people debunk them. And though we face a great challenge, success will make possible the fulfillment of great dreams. Great hopes and fears may stir enough people to enable the human race to win through.

Passionate concern and action will not be enough; we will also need sound policies. This will require more than good intentions and clear goals: we must also trace the factual connections in the world that will relate what we *do* to what we *get*. As we approach a technological crisis of unprecedented complexity, it makes sense to try to improve our institutions for judging important technical facts. How else can we guide the leading force and minimize the threat of terminal incompetence?

Institutions evolve. To evolve better fact-finding institutions, we can copy, adapt, and extend our past successes. These include the free press, the scientific community, and the courts. All have their virtues, and some of these virtues can be combined.

13

Finding the Facts

Fear cannot be banished, but it can be calm and without panic; and it can be mitigated by reason and evaluation.

—VANNEVAR BUSH

SOCIETY NEEDS BETTER WAYS to understand technology—this has long been obvious. The challenges ahead simply make our need more urgent.

The promise of technology lures us onward, and the pressure of competition makes stopping virtually impossible. As the technology race quickens, new developments sweep toward us faster, and a fatal mistake grows more likely. We need to strike a better balance between our foresight and our rate of advance. We cannot do much to slow the growth of technology, but we can speed the growth of foresight. And with better foresight, we will have a better chance to steer the technology race in safe directions.

Various approaches to guiding technology have been suggested. "The people must control technology" is a plausible slogan, but it has two possible meanings. If it means that we must make technology serve human needs, then it makes good sense. But if it means that the people as a whole must make technical decisions, then it makes very little sense. The electorate cannot judge the intricate links between technology, economy, environment, and life; people lack the needed knowledge. The people themselves agree: according to a U.S. National Science Foundation survey, 85 percent of U.S. adults believe that most citizens lack the knowledge needed to choose which technologies to develop. The public generally leaves technical judgments to technical experts.

Unfortunately, leaving judgment to experts causes problems. In *Advice and Dissent*, Primack and von Hippel point out that "to the extent that the Administration can succeed in keeping unfavorable information quiet and the public confused, the public welfare can be sacrificed with impunity to bureaucratic convenience and private gain." Regulators suffer more criticism when a new drug causes a single death than they do when the *absence* of a new drug causes a thousand deaths. They misregulate accordingly. Military bureaucrats have a vested interest in spending money, hiding mistakes, and continuing their projects. They mismanage accordingly. This sort of problem is so basic and natural that more examples are hardly needed. Everywhere, secrecy and fog make bureaucrats more comfortable; everywhere, personal convenience warps factual statements on matters of public concern. As technologies grow more complex and important, this pattern grows more dangerous.

Some authors consider rule by secretive technocrats to be virtually inevitable. In *Creating Alternative Futures*, Hazel Henderson argues that complex technologies "become *inherently* totalitarian" (her italics) because neither voters nor legislators can understand them. In *The Human Future Revisited*, Harrison Brown likewise argues that the temptation to bypass democratic processes in solving complex crises brings the danger "that if industrial civilization survives it will become increasingly totalitarian in nature." If this were so, it would likely mean our doom: we cannot stop the technology race, and a world of totalitarian states based on advanced technology needing neither workers nor soldiers might well discard most of the population.

Fortunately, democracy and liberty have met comparable challenges before. States grew too complex for direct democracy, but representative government evolved. State power threatened to crush liberty, but the rule of law evolved. Technology has grown complex, but this gives us no reason to ignore the people, discard the law, and hail a dictator. We need ways to handle technical complexity in a democratic framework, using experts as instruments to clarify our vision without giving them control of our lives. But technical experts today are mired in a system of partisan feuding.

A MESS OF EXPERTS

Government and industry—and their critics—commonly appoint expert committees that meet in secret, if they meet at all. These committees claim credibility based on *who they are*, not on *how they work*. Opposed groups recruit opposed Nobel laureates.

To gain influence in our mass democracy, groups try to outshout one another. When their views have corporate appeal, they take them to the public through advertising campaigns. When their views have pork-barrel appeal, they take them to legislatures through lobbying. When their views have dramatic appeal, they take them to the public through media campaigns. Groups promote their pet experts, the battle goes public, and quiet scientists and engineers are drowned in the clamor.

As the public conflict grows, people come to doubt expert pronouncements. They judge statements the obvious way, by their source. ("Of course she claims oil spills are harmless—she works for Exxon." "Of course he says Exxon lies—he works for Nader.")

When established experts lose credibility, demagogues can join the battle on an equal footing. Reporters—eager for controversy, striving for fairness, and seldom guided by technical backgrounds—carry all sides straight to the public. Cautious statements by scrupulous scientists make little impression; other scientists see no choice but to adopt the demagogues' style. Debates become sharp and angry, divisions grow, and the smoke of battle obscures the facts. Paralysis or folly often follows.

Our greatest problem is how we handle problems. Debates rage over the safety of nuclear power, coal power, and chemical wastes. Well-meaning groups backed by impressive experts clash again and again over dull, technical facts—dull, that is, save for their importance: What are the effects of low-level radiation, and how likely is a reactor meltdown? What are the causes and effects of acid rain? How well could space-based defenses block missile attacks? Do five cases of leukemia within three miles of a particular waste dump show a deadly hazard, or merely the workings of chance?

Greater issues lie ahead: How safe is this replicator? Will this active shield system be safe and secure? Will this biostasis procedure be reversible? Can we trust this AI system?

Disputes about technical facts feed broader disputes about policy. People may have differing values (which would you rather have, encephalitis or pesticide poisoning?) but their views of relevant facts often differ still more. (How often do *these* mosquitoes carry encephalitis? How toxic is *this* pesticide?) When different views of boring facts lead to disagreements about important policies, people may wonder, "How can they oppose us on this vital issue unless they have bad motives?" Disputes over facts can thus turn potential allies against one another. This hampers our efforts to understand and solve our problems.

People have disputed facts for millennia; only the prominence of technical disputes is new. Societies have evolved methods for judging facts about people. These methods suggest how we might judge facts about technology.

FROM FEUDS TO DUE PROCESS

Throughout history, groups have evolved ways to resolve disputes; the alternative has been feuds, open-ended and often deadly. Medieval Europeans used several procedures, all better than endless feuding:

They used trial by battle: opponents fought, and the law vindicated the victor.

They used compurgation: neighbors swore to the honesty of the accused; if enough swore, the charges were dropped.

They used trial by ordeal: in one, the accused was bound and thrown in a river; those who sank were innocent, those who floated, guilty.

They used judgment by secretive committees: the king's councilors would meet to judge and pass sentence as seemed fit. In England, they met in a room called the Star Chamber.

These methods supposedly determined *who did what*—the facts about human events—but all had serious shortcomings. Today we use similar methods to determine *what causes what*—the facts about science and technology:

We use trial by combat in the press: opponents fling sharp words until one side's case suffers political death. Unfortunately, this often resembles an endless feud.

We use compurgation: experts swear to certain facts; if enough swear the same, the facts are declared true.

We use judgment by secretive committees: selected experts meet to judge facts and recommend such actions as seem fit. In the United States, they often meet in committees of the National Academy of Sciences.

Trial by ordeal has passed from fashion, but combat in the press may well seem like torture to the quiet scientist with self-respect.

The English abolished Star Chamber proceedings in 1641, and they counted this a great achievement. Trial by combat, compurgation, and ordeal have likewise become history. We now value due process, at least when judging people.

Court procedures illustrate the principles of due process: Allegations must be specific. Both sides must have a chance to speak and confront each other, to rebut and cross-examine. The process must be public, to prevent hidden rot. Debate must proceed before a jury that both sides accept as impartial. Finally, a judge must referee the process and enforce the rules.

To see the value of due process, imagine its opposite: a process trampling all these principles would give one side a say and the other no chance to cross-examine or respond. It would meet in secret, allow vague smears, and lack a judge to enforce whatever rules might remain. Jurors would arrive with their decisions made. In short, it would resemble a lynch mob meeting in a locked barn—or a rigged committee drafting a report.

Experience shows the value of due process in judging facts about people; might it also be of value in judging facts about science and technology? Due process is a basic idea, not restricted to courts of law. Some AI researchers, for example, are building due-process principles into their computer programs. It seems that due process should be of use in judging technical facts.

In fact, the scientific literature—the chief forum of science—already embodies a form of due process: In good journals, scientific statements must be specific. In theory, given enough time and persistence, all sides may state their views in a dispute, since journals stand open to controversy. Though opponents may not meet face to face, they confront each other at a distance; they question and respond in slow motion, through letters and articles. Referees, like juries, evaluate evidence and reasoning. Editors, like judges, enforce rules of procedure. Publication keeps the debate open to public scrutiny.

In both journals and courts, conflicting ideas are pitted against one another under rules evolved to ensure a fair, orderly battle. These rules sometimes fail because they are broken or inadequate, but they are the best we have developed. Imperfect due process has proved better than no due process at all.

Why do scientists value refereed journals? Not because they trust all refereed journals, or trust everything printed in any one of them. Even the best due-process system won't grind out a stream of pure truth. Rather, they value refereed journals because they tend to reflect sound critical discussion. Indeed, they must: because journals compete with one another for papers, prestige, and readership, the best journals must be good indeed. Journals grind slowly, yet after enough rounds of publication and criticism they often grind out consensus.

Experience proves the value of both courts and journals. Their underlying similarity suggests that their value stems from a common source—due process. Due process can fail, but it is still the best approach known for finding the facts.

Today, courts and journals are not enough. Vital technical disputes go on and on because we have no rapid, orderly way to bring out the facts (and to delineate our ignorance). Courts are not suited to deal with technical questions. Journals are better, but they still have shortcomings. They took shape in a time of lower technology and slower advance, evolving to fit the limits of printing, the speed of mail, and the needs of academic science. But today, in a time when we desperately need better and swifter technical judgment, we find ourselves in a world that has telephones, jets, copiers, and express and electronic mail. Can we use modern technologies to speed technical debate?

Of course: scientists already use several approaches. Jets bring scientists from around the globe to conferences where papers are presented and discussed. But conferences handle controversy poorly: public decorum and tight schedules limit the vigor and depth of debate.

Scientists also join informal research networks linked by telephone, mail, computers, and copying machines; these accelerate exchange and discussion. But they are essentially private institutions. They, too, fail to provide a credible, public procedure for thrashing out differences.

Conferences, journals, and informal networks share some similar limitations. They typically focus on technical questions of scientific importance, rather than on technical questions of public-policy importance. Moreover, they typically focus on *scientific* questions. Journals tend to slight technological questions that lack intrinsic scientific interest; they often treat them as news items not worthy of checking by referees. Further, our present institutions lack any balanced way to present knowledge when it is still tangled in controversy. Though scientific review articles often present and weigh several sides, they do so from a single author's point of view.

All these shortcomings share a common source: scientific institutions evolved to advance science, not to sift facts for policymakers. These institutions serve their purpose well enough, but they serve other purposes poorly. Though this is no real failing, it does leave a real need.

AN APPROACH

We need better procedures for debating technical facts—procedures that are open, credible, and focused on finding the facts we need to formulate sound policies. We can begin by copying aspects of other due-process procedures; we then can modify and refine them in light of experience. Using modern communications and transportation, we can develop a focused, streamlined, journal-like process to speed public debate on crucial facts; this seems half the job. The other half requires distilling the results of the debate into a balanced picture of our state of knowledge (and by the same token, of our state of ignorance). Here, procedures somewhat like those of courts seem useful.

Since the procedure (a *fact forum*) is intended to summarize facts, each side will begin by stating what it sees as the key facts and listing them in order of importance. Discussion will begin with the statements that head each side's list. Through rounds of argument, cross-examination, and negotiation the *referee* will seek agreed-upon statements. Where disagreements remain, a *technical panel* will then write opinions, outlining what seems to be known and what still seems uncertain. The output of the fact forum will include background arguments, statements of agreement, and the panel's opinions. It might resemble a set of journal articles capped by a concise review article—one limited to factual statements, free of recommendations for policy.

This procedure must differ from that of a court in various ways. For example, the technical panel—the forum's "jury"—must be technically competent. Bias might lead a panel to misjudge facts, but technical incompetence would do equal harm. For this reason, the "jury" of a fact forum must be selected in a way that might be dangerous if allowed in courts of law. Since courts wield the power of the police, we use juries selected from the people as a whole to guard our liberty. This forces the government to seek approval from a group of citizens before it punishes someone, thus tying the government's actions to community standards. A fact forum, however, will neither punish people nor make public policy. The public will be free to watch the process and decide whether to believe its results. This will give people control enough.

Still, to make a fact forum fair and effective, we will need a good panel-selection procedure. Technical panels will correspond roughly to the expert committees appointed by governments or to the referees appointed by journals. To ensure fairness, a panel must be selected not by a committee, a politician, or a bureaucrat, but by a process that involves the consent of both sides in the dispute. In court proceedings, advocates can challenge and reject any jurors who seem biased; we can use a similar process in selecting the panel for a fact forum.

Experts who are directly involved in a dispute can't serve on the panel—they would either bias the panel or split it. The group sponsoring a fact forum must seek panelists who are knowledgeable in related fields. This seems practical because the methods of technical judgment (often based on experiments and calculations) are quite general. Panelists familiar with the fundamentals of a field will be able to judge the detailed arguments made by each side's specialists.

Other parts of the fact forum will also resemble those of courts and journals. A committee like a journal's editorial group will nominate a referee and panelists for a dispute. Advocates for each side, like authors or attorneys, will assemble and present the strongest case they can.

Despite these similarities, a fact forum will differ from a court: It will focus on technical questions. It will suggest no actions. It will lack government power. It will follow technical rules of evidence and argument. It will differ in endless details of tone and procedure. The analogy with a court is just that—an analogy, a source for ideas.

A fact forum will also differ from a journal: It will move as fast as mail, meetings, and electronic messages permit, rather than delaying exchanges by many months, as in typical journal publication. It will be convened around an issue, rather than being established to cover a scientific field. It will summarize knowledge to aid decisions, rather than serving as a primary source of data for the scientific community. Although a series of fact forums won't replace a journal, they will help us find and publicize facts that could save our lives.

Dr. Arthur Kantrowitz (a member of the National Academy of Sciences who is accomplished in fields ranging from medical technology to high-power lasers) originated the concept I have just outlined. He at first called it a "board of technical inquiry." Journalists promptly dubbed it a "science court." I have called it a "fact forum"; I will reserve the term "science court" for a fact forum used (or proposed) as a government institution. Proposals for due process in technical disputes are still in flux; different discussions use different terms.

Dr. Kantrowitz's concern with due process arose out of the U.S. decision to build giant rockets to reach the Moon in one great leap; he, backed by the findings of an expert committee, had recommended that NASA use several smaller rockets to carry components into a low orbit, then plug them together to build a vehicle to reach the Moon. This approach promised to save billions of dollars and develop useful space-construction capabilities as well. No one answered his arguments, yet he failed to win his case. Minds were set, politicians were committed, the report was locked in a White House safe, and the debate was closed. The technical facts were quietly suppressed in the interests of those who wanted to build a new generation of giant rockets.

This showed a grave flaw in our institutions—one that persists, wasting our money and increasing the risk of a disastrous error. Dr. Kantrowitz soon reached the now obvious conclusion: we need due-process institutions for airing technical controversies.

Dr. Kantrowitz pursued this goal (in its science-court form) through discussions, writings, studies, and conferences. He won endorsements for the science court idea from Ford, Carter, and Reagan —as candidates. As Presidents, they did nothing, though a presidential advisory task force during the Ford administration did detail a proposed procedure.

Still, progress has been made. Although I have used the future tense in describing the fact forum, experiments have begun. But before describing a path to due process, it makes sense to consider some of the objections.

WHY NOT DUE PROCESS?

Critics of this idea (at least in its science court version) have often disagreed with one another. Some have objected that factual disputes are unimportant, or that they can be smoothed over behind closed doors; others have objected that factual disputes are too deep and important for due process to help. Some have warned that science courts would be dangerous; others have warned that they would be impotent. These criticisms all have some validity: due process will be no cure-all. Sometimes it won't be needed, and sometimes it will be abused. Still, one might equally well reject penicillin on the grounds that it is sometimes ineffective, unnecessary, or harmful.

These critics propose no alternatives, and they seldom argue that we have due process today, or that due process is worthless. We must deal with complex, technical issues on which millions of lives depend; dare we leave these issues to secretive committees, sluggish journals, media battles, and the technical judgment of politicians? If we distrust experts, should we accept the judgment of secretive committees appointed in secret, or demand a more open process? Finally, can we with our present system cope with a global technology race in nanotechnology and artificial intelligence?

Open, due-process institutions seem vital. By letting all sides participate, they will harness the energy of conflict to a search for the facts. By limiting experts to describing the facts, they will help us cope with technology without surrendering our decisions to technocrats. Individuals, companies, and elected officials will keep full control of policy; technical experts will still be able to recommend policies through other channels.

How can we distinguish facts from values? Karl Popper's standard seems useful: a statement is factual (whether true or false) if an experiment or observation could in principle disprove it. To some people, the idea of *examining facts* without considering values suggests the idea of *making policy* without considering values. This would be absurd: by their very nature, policy decisions will always involve both facts and values. Cause and effect are matters of fact, telling us what is possible. But policy also involves our values, our motives for action. Without accurate facts, we won't get the results we seek, but without values—without desires and preferences—we wouldn't seek anything in the first place. A process that uncovers facts can help people choose policies that will serve their values.

Critics have worried that a science court will (in effect) declare the Earth to be flat, and then ignore an Aristarchus of Samos or a Magellan when he finds evidence to the contrary. Errors, bias, and imperfect knowledge will surely cause some memorable mistakes. But members of a technical panel need not claim a bogus certainty. They can instead describe our knowledge and outline our ignorance, sometimes stating that we simply don't know, or that present evidence gives only a rough idea of the facts. This way, they will protect their reputations for good judgment. When new evidence arrives, a question can be reopened; ideas need no protection from double jeopardy.

If fact forums become popular and respected, they will gain influence. Their success will then make them harder to abuse: many competing groups will sponsor them, and a group that abuses the procedure will tend to gain a bad reputation and be ignored. No single sponsoring group will be able to obscure the facts about an important issue, if fact forums gain reliability through redundancy.

No institution will be able to eliminate corruption and error, but fact forums will be guided, however imperfectly, by an improved standard for the conduct of public debate; they will have to fall a long way before becoming worse than the system we have today. The basic case for fact forums is that (1) due process is the right approach to try, and (2) we will do better if we try than if we don't.

BUILDING DUE PROCESS

Anthropologist Margaret Mead was invited to a colloquium on the science court to speak against the idea. But when the time came, she spoke in its favor, remarking that "We need a new institution. There isn't any doubt about that. The institutions we have are totally unsatisfactory. In many cases, they are not only unsatisfactory, they involve a prostitution of science and a prostitution of the decision-making process." People with no vested interest in the existing institutions often agree with her evaluation.

If finding the facts about technology really is crucial to our survival, and if due process really is the key to finding the facts, then what can we do about it? We needn't begin with perfect procedures;

we can begin with informal attempts to improve on the procedures we have. We can then evolve better procedures by varying our methods and selecting those that work best. Due process is a matter of degree.

Existing institutions could move toward due process by modifying some of their rules and traditions. For example, government agencies could regularly consult opposing sides before appointing the members of an expert committee. They could guarantee each side the right to present evidence, examine evidence, and cross-examine experts. They could open their proceedings to observers. Each of these steps would strengthen due process, changing Star Chamber proceedings into institutions more worthy of respect.

The public benefits of due process won't necessarily make it popular among the groups being asked to change, however. We haven't heard the thunder of interest groups rushing to test their claims, nor the cries of joy from committees as they throw open their doors and submit to the discipline of due process. Nor have we heard reports of politicians renouncing the use of spurious facts to hide the political basis of their decisions.

Yet three U.S. presidential candidates *did* endorse science courts. The Committee of Scientific Society Presidents, which includes twenty-eight of the leading scientific societies in the U.S., also endorsed the idea. The U.S. Department of Energy used a "science court-like procedure" to evaluate competing fusion-power proposals, and declared it efficient and useful. Dr. John C. Bailar of the National Cancer Institute, after failing to make medical organizations recognize the dangers of X rays and reduce their use in mass screening, proposed holding a science court on the subject. His opponents then backed down and changed their policies—apparently, the mere threat of due process is already saving lives. Nevertheless, the old ways continue almost unchallenged.

Why is this? In part, because knowledge is power, and hence jealously guarded. In part, because powerful groups can readily imagine how due process would inconvenience them. In part, because an effort to improve problem-solving methods lacks the drama of a campaign to fight problems directly; a thousand activists bail out the ship of state for every one who tries to plug the holes in its hull.

Governments may yet act to establish science courts, and any steps they may take toward due process merit support. Yet it is reasonable to fear government sponsorship of science courts: centralized power tends to beget lumbering monsters. A central "Science Court Agency" might work well, do little visible harm, and yet impose a great hidden cost: its very existence (and the lack of competition) might block evolutionary improvements.

Other paths lie open. Fact forums will be able to exert influence without help from special legal powers. To have a powerful effect, their results only need to be more credible than the assertions of any particular person, committee, corporation, or interest group. A well-run fact forum will carry credibility in its very structure. Such forums could be sponsored by universities.

In fact, Dr. Kantrowitz recently conducted an experimental procedure at the University of California at Berkeley. It centered around public disputes between geneticist Beverly Paigen and biochemist William Havender regarding birth defects and genetic hazards at the Love Canal chemical dump site. They served as advocates; graduate students served as a technical panel. The procedure involved meetings spread over several weeks, chiefly spent in discussing areas of agreement and disagreement; it wound up with several sessions of public cross-examination before the panel. Both the advocates and the panel agreed to eleven statements of fact, and they clarified their remaining disagreements and uncertainties.

Arthur Kantrowitz and Roger Masters note that "in contrast to the difficulties experienced in the many attempts to implement a science court under government auspices, encouraging results were obtained in the first serious attempt . . . in a university setting." They remark that the traditions and resources of universities make them natural settings for such efforts. They plan more such experiments.

This shows a decentralized way to develop due-process institutions, one that will let us outflank existing bureaucracies and entrenched interests. In doing this, we can build on established principles and test variations in the best evolutionary tradition.

Leaders concerned with a thousand different issues—even leaders on opposing sides of an issue—can join in support of due process. In *Getting to Yes*, Roger Fisher and William Ury of the Harvard Negotiation Project point out that opponents tend to favor impartial arbitration when each side believes itself right. Both sides expect to win, hence both agree to participate. Fact forums should attract honest advocates and repel charlatans.

As due process becomes standard, advocates with sound cases will gain an advantage even if their opponents refuse to cooperate—"If they won't defend their case in public, why listen to them?" Further, many disputes will be resolved (or *avoided)* without the trouble of actual proceedings: the prospect of a refereed public debate will encourage advocates to check their facts before they take stands. Establishing fact forums could easily do more for a sound cause than would the recruitment of a million supporters.

Yet today well-meaning leaders may feel forced to exaggerate their cases simply to be heard above the roar of their opponents' press releases. Should they regretfully oppose the discipline of fact forums? Surely not; due process can heal the social illness that forced them away from the facts in the first place. By making their points in a fact forum, they can recover the self-respect that comes with intellectual honesty.

Unearthing the truth may undermine a cherished position, but this cannot harm the interests of a true public-interest group. If one must move a bit to make room for the truth, so what? Great leaders have shifted positions for worse reasons, and due process will make opponents shift too.

Gregory Bateson once stated that "no organism can afford to be conscious of matters with which it could deal at unconscious levels." In the organism of a democracy, the conscious level roughly corresponds to debate in the mass media. The unconscious levels consist of whatever processes ordinarily work well enough without a public hue and cry. In the media, as in human consciousness, one concern tends to drive out another. This is what makes conscious attention so scarce and precious. Our society needs to identify the facts of its situation more swiftly and reliably, with fewer distracting feuds in the media. This will free public debate for its proper task—judging procedures for finding facts, deciding what we want, and helping us choose a path toward a world worth living in.

As change quickens, our need grows. To judge the risk of a replicator or the feasibility of an active shield, we will need better ways to debate the facts. Purely social inventions like the fact forum will help, but new social technologies based on computers also hold promise. They, too, will extend due process.

14

The Network of Knowledge

Computers . . . have penetrated our daily lives and are becoming society's central nervous system.

—TOHRU MOTO-OKA

TO PREPARE for the assembler breakthrough, society must learn to learn faster. Fact forums will help, but new technologies may help even more. With them, we will be able to spread, refine, and combine our information far faster than ever before.

Information overload has become a well-known problem: pieces of knowledge pile up too fast for people to sort and make sense of them. Thousands of technical journals cover thousands of subjects. Published articles mount by over a million per year. Fact forums will help us to clear away falsehoods, which will smooth our efforts to make sense of the world. But any such formal institution will be overwhelmed by the modern flood of information: fact forums will be able to deal with only a fraction—though an important fraction—of the facts, and they will inevitably be somewhat sluggish. Formal institutions can tap only a tiny fraction of the mental energies of our society.

Today, our information systems hamper our progress. To see the problem, imagine handling a piece of information: You uncovered it —how do you spread it? Someone else published it—how do you find it? You found it—where do you file it? You see an error—how do you correct it? Your file grows—how do you organize it?

We now handle information clumsily. Our traditional electronic media are vivid and entertaining, but they are ill suited to handling complex, long-term debates; how could you, as a viewer, file, organize, or correct the information in a television documentary? In short, how could you make it a well-integrated part of an evolving body of knowledge? We can handle complex debates better using paper media, yet the weeks (or *years*) of delay in a typical publication process slow debate to a crawl. And even paper publications are difficult to file, organize, and correct. Printers produce bundles of inked paper; through heroic efforts, librarians and scholars manage to link and organize them in a loose fashion. Yet indexes, references, and corrections simply add more pages or more editions, and tracing the links they represent remains tedious.

Books and other paper bundles work, after a fashion. They hold many of our cultural treasures, and we now have no better way to publish most things. Still, they leave great room for improvement.

Our trouble in spreading, correcting, and organizing information leaves our shared knowledge relatively scarce, incorrect, and disorganized. Because established knowledge is often hard to find, we often do without it, making us seem more ignorant than we need be. Can new technologies help us?

They have before. The invention of the printing press brought great advances; computer-based text services promise yet more. To see how our information systems could be better, though, it may help to see how they could be worse. Consider, then, an imaginary mess and an imaginary solution:

The Tale of the Temple

Once upon a time, there lived a people with an information problem. Though they had replaced their bulky clay tablets with paper, they used it oddly. In the heart of their land stood a stately dome. Beneath the dome lay their great Chamber of Writings. Within this chamber lay a broad mound of paper scraps, each the size of a child's hand.

From time to time, a scholar would journey to this temple of learning to offer knowledge. A council of scribes would judge its worth. If it proved worthy, they would inscribe it on a scrap of paper and ceremoniously fling it upon the heap.

From time to time, some industrious scholar would come to seek knowledge—to rummage through the heap in search of the needed scrap. Some, skilled in such research, could find a particular scrap in as little as a month. The scribes always welcomed researchers: they were so rare.

We moderns can see their problem: in a disorderly heap, each added scrap buries the rest (as on so many desks). Every scrap is separate, unrelated to the others, and adding references would provide little help when finding a scrap takes months. If we used such a heap to store information, our massive, detailed writings on science and technology would become almost useless. Searches would take years, or lifetimes.

We moderns have a simple solution: we place pages in order. We place page after page to make a book, book after book to fill a shelf, then fill a building with shelves to make a library. With pages in order, we can find them and follow references more rapidly. If the scribes employed scholars to stack scraps by subject, their research would grow easier.

Yet, when faced with stacks on history, geography, and medicine, where should the scholars put scraps on historical geography, geographical epidemiology, and medical history? Where should they put scraps on "The History of the Spread of the Great Plague"?

But in our imaginary land, the scribes choose another solution: they send for a magician. But first they turn scholars loose in the chamber with needles and thread to run strands from scrap to scrap. Thread of one color links a scrap to the next in a series, another color leads to a reference, another to a critical note, and so forth. The scholars weave a network of relationships, represented by a network of threads. At last, the magician (with flashing eyes and floating hair) chants a spell, and the whole mess heaves slowly into the air to float like a cloud in the dome. Ever after, a scholar holding a scrap need only tap a thread knotted to its edge to make the linked scrap leap to hand. And the threads, magically, never tangle.

Now the scholars can link scraps on "The History of the Spread of the Great Plague" to related scraps on history, geography, and medicine. They can add all the notes and texts they please, linking them to best advantage. They can add special index scraps, able to bring instantly to hand whatever they list. They can place links wherever they wish, weaving a network of knowledge to match the connections of the real world.

We, with our inert stacks of paper, could only envy them—if we didn't have computers.

MAGIC PAPER MADE REAL

In 1945, Vannevar Bush proposed a system he called a "memex." It was to be a desk-sized device, crammed with microfilm and mechanisms, able to display stored pages and let the user note relationships among them. A microfilm memex was never built, but the dream lived on.

Today, computers and screens are becoming cheap enough to use for ordinary reading and writing. Some paper publishers have become electronic publishers, making magazines, newspapers, and journals available through computer networks. And with the right programs, text-handling computers will let us link this information in ways even better than magic thread.

Theodor Nelson, the originator of the idea, has dubbed the result "hypertext": text linked in many directions, not just in a one-dimensional sequence. Readers, authors, and editors using a hypertext system will generally ignore the workings of its computers and screens just as they have generally ignored the mechanics of photocomposition and offset lithography in the past. A hypertext system will simply act like magic paper; anyone who fiddles with it will soon become familiar with its basic abilities. Still, a description of one system's structure will help in explaining how hypertext will work.

In the approach followed by the Xanadu hypertext group (in San Jose, California), the core of the system—the back end—is a computer network able to store both documents and links between documents. An initial system might be a single-user desktop machine; eventually a growing network of machines will be able to serve as an electronic library. Stored documents will be able to represent almost anything, whether novels, diagrams, textbooks, or programs—eventually, even music or movies.

Users will be able to link any part of any document to any other. When a reader points to one end of a link (whether it is shown on the screen by underlining, an asterisk, or a picture of colored thread), the system will fetch and display the material at the other end. Further, it will record new versions of a large document without storing additional copies; it need store only the parts that are changed. This will let it inexpensively store the earlier versions of any document published and modified on the system. It will do all this rapidly, even when the total amount of information stored becomes immense. A network of such machines could eventually mature into a world electronic library.

To locate material in most computer-based text systems, the user must supply key words or obscure codes. Hypertext, too, will be able to link text to codes, to key words, or even to a simulated card catalog, but most readers will probably prefer just to read and point to links. As Theodor Nelson has remarked, hypertext will be "a new form of reading and writing, in a way just like the old, with quotations and marginalia and citations. Yet it will also be socially self-constructing into a vast new traversable framework, a new literature."

What the reader will see when browsing through this framework will depend partly on the reader's own part of the system, the "front-end" machine, perhaps a personal computer. The back end will just file and fetch documents; the front end will order them fetched at the reader's request and will display them to suit the reader's taste.

To imagine how this will appear to a user, picture a screen the size of this open book, covered with print the size you are reading now—clear print, on a good screen. Today the screen would resemble a television set, but within a few years it could be a booklike, lap-sized object with a cord to an information outlet. (With nanotechnology, we can eliminate the cord: a book-sized object will be able to hold a hypertext system containing images of every page in every book in the world, stored in fast, molecular-tape memory.)

In *this* book—the one now in your hands—I could describe Theodor Nelson's books about hypertext, *Literary Machines* and *Computer Lib,* but you couldn't see them on these pages. Their pages are elsewhere, leaving you trapped for the moment in one author's writings. But if this were a hypertext system and I, or someone else, had added the obvious link, you could point to the words *"Literary Machines"* here, and a moment later the text on the opposite page would clear, to be replaced by Ted Nelson's table of contents or by my selection of his quotes. From there, you could step into his book and roam through it, perhaps while having your front-end system display any notes that I had linked to his text. You could then return here (perhaps now displaying *his* notes on *my* text) or move on to yet other documents linked to his. Without leaving your chair, you could survey all the major writings on hypertext, moving from link to link through any number of documents.

By keeping track of links (say, between outlines, drafts, and reference materials), hypertext will help people write and edit more ambitious works. Using hypertext links, we can weave our knowledge into coherent wholes. John Muir observed that "When we try to pick out anything by itself, we find it hitched to everything else in the universe." Hypertext will help us keep ideas hitched together in ways that better represent reality.

With hypertext, we will be better able to gather and organize knowledge, increasing our effective intelligence. But for information gathering to be effective, it must be decentralized; information scattered among many minds cannot easily be put into the system by a few specialists. The Xanadu group proposes a simple solution: let everyone write, and have the system automatically pay royalties to the authors whenever readers use their material. Publishing will be easy and people will be rewarded for providing what other people want.

Imagine what you yourself have wanted to say about ideas and events. Imagine the insightful comments that are even now fading from the memories of both speakers and listeners all over the world. On a hypertext system, comments will be easy to publish and easy to find. Imagine the *questions* that have bothered you. You could publish them, too; someone finding an answer could then publish a reply.

Since everyone on the system will be able to write text and links, the hypertext network will accumulate great stores of knowledge and wisdom and even greater heaps of utter garbage. Hypertext will include old news, advertising, graffiti, ranting, and lies—so how will a reader be able to avoid the bad and focus on the good? We could appoint a central editorial committee, but this would destroy the openness of the system. Sorting information is itself an information problem for which hypertext fortunately will help us evolve good solutions.

Since hypertext will be able to do almost everything that paper systems can, we can at least use the solutions we already have. Publishers have established reputations in the paper-text media, and many of them have begun to move into electronic publishing. On a hypertext system they will be able to publish on-line documents that meet their established standards. Readers so inclined will be able to set their front-end systems to display only these documents, automatically ignoring the new garbage. To them, the hypertext system will seem to contain only material by established publishers, but material made more available by electronic distribution and by hypertext links and indexes. True garbage will still be there (so long as its authors have paid a small storage fee for their material), yet garbage need not intrude on any reader's screen.

But we will be able to do better than this. Approval of a document (shown by links and recommendations) can come from anyone; readers will pay attention to material recommended by whomever they respect. Conversely, readers who find documents they like will be able to see who has recommended them; this will lead readers to discover people who share their interests and concerns. Indirectly, hypertext will link people and speed the growth of communities.

When publishing becomes so fast and easy, writers will produce more material. Since hypertext will encourage free-lance editing, editors will find themselves with more work to do. Documents that quote, list, and link other documents will serve as anthologies, journals, or instant-access indexes. The incentive of royalties will encourage people to help readers find what they want. Competing guides to the literature will swiftly appear—and guides to the guides.

Hypertext links will be better than paper references, and not merely in speed. Paper references let an industrious reader follow links *from* one document *to* another—but try finding which documents refer *to* one you are reading! Today, finding such references requires the cumbersome apparatus of a citation index, available only in research libraries, covering limited topics, and months out of date. Hypertext links will work in *both* directions, letting readers find comments on what they are reading. This means a breakthrough: it will subject ideas to more thorough criticism, making them evolve faster.

The evolution of knowledge—whether in philosophy, politics, science, or engineering—requires the generation, spread, and testing of memes. Hypertext will speed this process. Paper media handle the process of generation and spread fairly well, but they handle testing clumsily.

Once a bad idea reaches print, it takes on a life of its own, and even its author can seldom drive a stake through its heart. A devastating refutation of the bad idea becomes just another publication, another scrap of paper. Days or years later, readers who encounter the bogus idea will still be unlikely to have chanced upon its refutation. Thus, nonsense lives on and on. Only with the advent of hypertext will critics be able to plant their barbs firmly in the meat of their targets. Only with hypertext will authors be able to retract their errors, not by burning all the libraries or by mounting a massive publicity campaign, but by revising their text and labeling the old version "retracted." Authors will be able to eat their words quietly; this will give them some compensation for the fiercer criticism.

Critics will use clear refutations to skewer nonsense (such as false limits to growth), clearing it from the intellectual arena—though not from the record—almost as soon as it pops into sight. Guides to good criticism will help readers see whether an idea has survived the worst objections yet raised. Today, the absence of known criticism doesn't mean much, because brief critical comments are hard to publish and hard to find. In an established hypertext system, though, ideas that have survived all known criticism will have survived a real challenge. They will gain real and growing credibility.

LINKING OUR KNOWLEDGE

The advantages of hypertext run deep; this is why they will be great. Hypertext will let us represent knowledge in a more natural way. Human knowledge forms an unbroken web, and human problems sprawl across the fuzzy boundaries between fields. Neat rows of books do a poor job of representing the structure of our knowledge. Librarians have labored to make these rows more like nets by inventing better ways to index, reference, and arrange pieces of paper. Yet despite the noble efforts and victories of librarians, library research still daunts all but a dedicated minority of the reading public. Libraries have evolved toward hypertext, yet the mechanics of paper still hobbles them. Hypertext systems will let us take a giant step in a direction we have been moving since the invention of writing.

Our very memories work through associations, through links that make recollections recallable. AI workers also find associations essential to making knowledge useful; they program what they call "semantic nets" to build knowledge representation systems. On paper, associations among words make a thesaurus useful; in the mind, one's working vocabulary relies on fast, flexible associations among words. Indeed, relationships in memory supply the context that gives meaning to our ideas. Using hypertext, people will associate ideas through published links, enriching their meaning and making them more available—indeed, making them more like parts of our own minds.

When we change our minds about what the world is like and where it is going, we change our internal networks of knowledge. Reasoned change often requires that we compare competing patterns of ideas.

To judge a worldview presented in a book, a reader must often remember or reread explanations from earlier pages—or from a conflicting article seen last year. But human memory is faulty, and digging around in old paper often seems like too much work. Knowing of this problem, authors vacillate between putting too much in (thus boring their readers) and leaving too much out (thus leaving weak spots in their discussions). Inevitably, they do both at once.

Hypertext readers will be able to see whether linked sources support an idea or linked criticisms explode it. Authors will write pithy, exciting summaries of ideas and link them to the lengthy, boring explanations. As authors expound and critics argue, they will lay out their competing worldview networks in parallel, point by point. Readers still won't be able to judge ideas instantly or perfectly, but they will be able to judge them faster and better. In this way, hypertext will help us with a great task of our time: judging what lies ahead, and adjusting our thinking to prospects that shake the foundations of established worldviews. Hypertext will strengthen our foresight.

By now, many useful applications of a mature hypertext system will be obvious—or as obvious as they can be today, before we have experienced them directly. Carrying news is one such obvious application.

News shapes our view of the world, but modern media sharply limit what reporters can portray. Often, stories about technology and world events only make sense in a broader context, but the limited space and onrushing deadlines of publishing strip needed context from stories. This weakens our grasp of events. Using hypertext, reporters will find it easy to link today's news to broader background discussions. What is more, the people in the stories and casual observers will be able to have their say, linking their comments to the reporter's story.

Advertising greases the wheels of the economy, leading (and misleading) us to available products. Well-informed consumers can avoid shoddy, overpriced goods, but the needed research and comparison shopping gobbles time. On a hypertext system, though, consumer service companies will assemble comparative catalogs, linking descriptions of competing products to one another, to test results, and to reports from consumers.

In education, we learn best when we are interested in what we read. But most books present ideas in just one sequence, at just one level of difficulty, regardless of a learner's background or interests. Again, popular demand will favor the growth of useful networks in hypertext. People will make links between similar presentations written at different levels. Students will be able to read at a comfortable level, peeking at parallel discussions that reach a bit deeper. Hard material will grow easier to handle, because links to primers and basic definitions will let readers pause for review—instantly, privately, and without embarrassment. Other links will lead in all directions to related material; links in a description of a coral reef will lead to both texts on reef ecology and tales of hungry sharks. When we can gratify momentary interests almost instantly, learning will become more fun. More people might then find it addictive.

Due process will thrive in hypertext. Because it will be open to all sides, and will allow questioning, response, and so forth, hypertext debate will have an inherent due-process quality. Indeed, hypertext will be an ideal medium for conducting fact forums. Forum procedures, in turn, will complement hypertext by distilling its wide-ranging debates into a clear (though tentative) statement of key technical facts.

In a final obvious effect, hypertext will reduce the problem of out-of-context quotes: readers will be able to make the original context reappear around any quote in the system at the touch of a button. This will be valuable, and not just to prevent misrepresentation of an author's position; indirect benefits may matter more. Reasonable statements torn loose from their background can seem absurd, but hypertext authors will know that "absurd" quotations will lead readers directly back to the author's original context. This will encourage bolder writing, giving memes based on evidence and reason an advantage over those based on mere convention and timidity.

Perhaps the most important (yet least vivid) benefit of hypertext will be a new ability to see *absences*. To survive the coming years, we must evaluate complex ideas correctly, and this requires judging whether an argument is full of holes. But today we have trouble seeing holes.

Still more difficult is recognizing the *absence* of fatal holes, yet this is the key to recognizing a sound argument. Hypertext will help us. Readers will scrutinize important arguments, attaching conspicuous objections where they find holes. These objections will make holes so consistently visible that an *absence* of good objections will clearly indicate an *absence* of known holes. It may be hard to appreciate how important this will be: the human mind tends not to recognize the problems caused by our inability to see the absence of holes, to say nothing of the opportunities this inability makes us miss.

For example, imagine that you have an idea and are trying to decide whether it is sound and worth publishing. If the idea isn't obvious, you might doubt its truth and not publish it. But if it does seem obvious, you might well assume that it has already been published, but that you just can't find out where. Hypertext, by making things much easier to find, will make it easier to see that something has *not* been published. By making holes in our knowledge more visible, hypertext will encourage hole-filling.

To understand and guide technology, we need to find the errors—including omissions—in complex technological proposals. Because we do this poorly, we make many mistakes, and the visibility of these mistakes makes our incompetence a vivid and menacing fact. This encourages prudence, yet it can also encourage paralysis: because we have difficulty seeing holes, we fear them everywhere, even where they do not exist. Hypertext will help build confidence, where confidence is justified, by exposing problems more reliably.

DANGERS OF HYPERTEXT

Like most useful tools, hypertext could be used to do harm. Though it will help us keep track of facts, it could also help governments keep track of us. Yet, on balance, it may serve liberty. Designed for decentralization—with many machines, many writers, many editors—hypertext may help citizens more than it helps those who would rule them. Government data banks are growing anyway. Hypertext systems might even help us keep an eye on them.

Relying on electronic publishing holds another danger. Governments in the United States and elsewhere have often interpreted the ideal once expressed as "freedom of speech" and "of the press" to mean only freedom to talk and to sell inked paper. Governments have regulated the use of radio and television, requiring them to serve a bureaucracy's shifting notion of the public interest. Practical limitations on the number of broadcasting channels once gave some excuse for this, but those excuses must stop here. We must extend the principles of free speech to new media.

We would be horrified if the government ordered agents into libraries to burn books. We should be equally horrified when the government seeks to erase public documents from electronic libraries. If hypertext is to carry our traditions, then what is published must remain so. An electronic library will be no less a library for its lack of shelves and paper. Erasure will make no flame and smoke, but the stench of book-burning will remain.

FROM DESKTOP TO WORLD LIBRARY

Some of the benefits I have described will only result from a large, highly evolved hypertext system—one already serving as a forum for broad debates and on its way to becoming a world electronic library. Such a system may not have time to mature before the assembler and AI breakthroughs arrive. For hypertext to get off the ground, small systems must have practical applications, and for hypertext to help us handle the technology race, small systems must have an effect beyond their size. Fortunately, we can expect substantial benefits almost from the outset.

Individual hypertext machines will be able to serve several users at once. Even without linking to anything in the outside world, they will help companies, associations, and research groups handle complex information.

Outside links will be easy, though. The number of publicly available data bases has grown from a few dozen in the mid-1960s to a few hundred in the mid-1970s to a few thousand in the mid-1980s. Companies have made these available through computer networks. Hypertext systems will be able to fetch material from these data bases, storing the access codes instead of the actual text. This information will only *seem* to be in the hypertext system, but that will be good enough for many purposes.

People will use early systems to provide a community with a dial-up service like that of existing computer bulletin boards but better. Special-interest discussion groups have already emerged on computer networks; they will find hypertext a better medium for exchanging information and views.

Early hypertext systems will also help us build and run organizations. Ordinary computer conferencing (simply sending short messages back and forth) already helps groups communicate. Advantages over face-to-face conferences include lower costs (no need to travel), smoother interactions (no need to wait or to interrupt people), and a better meeting of minds (through clearer messages and fewer personality clashes). Hypertext communications will extend these benefits by giving participants better tools for referencing, comparing, and summarizing. Because hypertext debates will need no single editor, they will allow organizations to become more open.

Using a hypertext service by telephone from a home computer during off-peak hours will probably cost several dollars per hour at first. This cost will fall over time. For several decades now, the real cost of computers has dropped by about a factor of ten every ten years; the cost of communications has also declined. Hypertext systems will be affordable to a substantial number of people almost as soon as they become available. Within ten years, the costs seem likely to fall low enough for mass-market use.

Electronic publishing is already catching on. The *Academic American Encyclopedia,* structured as a simple hypertext, has been made available to 90,000 subscribers, including 200 libraries and eight schools. *Time* magazine reports that children use it eagerly. Terminals in libraries already can access the text of scores of newspapers, magazines, and professional journals.

We won't need to wait for a universal system to enjoy universal benefits, because hypertext will begin to make a difference very early. We need hypertext in the hands of students, writers, researchers, and managers for the same reason that we need textbooks in schools, instruments in laboratories, and tools in workshops. Some books have made a great difference, even when read by fewer persons than one in a thousand, because they have sent new ideas rippling across society. Hypertext will do likewise, helping to refine ideas that will then spread more widely through the established print and broadcast media.

HYPERTEXT AND PRINTING PRESS

How will the hypertext revolution compare to the Gutenberg revolution? Some numbers suggest the answer.

Printing with movable type cut the cost of books dramatically. In the fourteenth century, the King's Advocate of France had only seventy-six books, yet this was considered a large library. Books embodied weeks of skilled labor—copyists were *literate*. The peasant masses could neither afford books nor read them.

Today, a year's labor can pay for thousands of books. Many homes hold hundreds; large libraries hold millions. Printing cut the cost of books by a hundredfold or more, setting the stage for mass literacy, mass education, and the ongoing world revolution of technology and democracy.

And hypertext? Gutenberg showed Europe how to arrange metal type to print pages; hypertext will let us rearrange stored text and send it cross-country at the speed of light. Printing put stacks of books in the home and mountains of books in libraries; hypertext will in effect bring these mountains of books to every terminal. Hypertext will extend the Gutenberg revolution by increasing the quantity of information available.

Yet its other advantages seem greater. Today, following a reference in a library typically takes minutes; with luck, a few hundred seconds—but it can take days or more, if the material is unpopular and hence absent, or too popular and hence missing. Hypertext will cut this delay from hundreds of seconds to about one second. Thus where the Gutenberg revolution reduced the labor cost of *producing* text by several hundredfold, the hypertext revolution will reduce the labor cost of *finding* text by several hundredfold. This will be a revolution indeed.

As I have discussed, making links more convenient will change the texture of text, bringing a revolution not merely in quantity but in quality. This increase in quality will take many forms. Better indexes will make information easy to find. Better critical discussion will weed out nonsense and help sound ideas thrive. Better presentation of wholes will highlight the holes in our knowledge.

With abundant, available, high-quality information, we will seem more intelligent. And this will increase our chances of handling the coming breakthroughs *right*. What could be more useful? Next time you see a lie being spread or a bad decision being made out of sheer ignorance, pause, and think of hypertext.

15

Worlds Enough, and Time

The difficulty lies, not in the new ideas, but in escaping the old ones, which ramify, for those brought up as most of us have been, into every corner of our minds.

—JOHN MAYNARD KEYNES

I HAVE DESCRIBED how advances in chemistry and biotechnology will lead to assemblers, which will bring nanocomputers, replicators, and cell repair machines. I have described how advances in software will lead to automated engineering and artificial intelligence. Together, these advances will make possible a future rich in possibilities, one of which is our own destruction. If we use fact forums and hypertext to strengthen our foresight, we may nonetheless avoid annihilation and move forward—but toward what?

Toward a worldwide transformation which can, if we succeed, bring abundance and long life to all who wish them. And this is a prospect that quite naturally stirs dreams of utopia.

A standard-issue utopia, as everyone knows, would be static, boring, and dreadful—in fact, it would be no utopia at all. Yet again and again utopian dreams have changed history, whether for good or ill. Dangerous dreams have led people to kill in the name of love, and to enslave in the name of brotherhood. All too often the dream has been impossible and the attempt to achieve it has been disastrous.

We need useful dreams to guide our actions. A useful dream must show us a possible, desirable goal, and steps toward that goal must bring positive results. To help us cooperate in guiding the technology race, we will need goals that appeal to people with differing dreams—but what goals could serve? It seems that they must hold room for diversity. Likewise, what goals chosen today, so near the dawn of intelligence, could prove worthy of the future's potential? It seems that they must hold room for progress.

Only one sort of future seems broad enough to have broad appeal: an open future of liberty, diversity, and peace. With room for the pursuit of many different dreams, an open future will appeal to many different people. Grander schemes, such as establishing a uniform world order, seem more dangerous. If "one world, or none" means imposing a single social system on a world of hostile nuclear powers, then it seems a recipe for disaster. "Many worlds, or none" seems our real choice, if we can develop active shields to secure peace.

We may be able to do so. Using automated engineering systems of the sort described in Chapter 5, we will be able to explore the limits of the possible at a million times human speed. We will thus be able to outline the ultimate limits to the technology race, including the arms race. With shields based on that knowledge, it seems that we would be able to secure a stable, durable peace.

Advancing technology need not push the world into any single mold. Many people once feared that ever larger machines and ever larger organizations would dominate our future, crushing diversity and human choice. Indeed, machines can grow bigger, and some may. Organizations can grow bigger, and some may. But stinking, clanking machines and huge bureaucracies have already begun to seem old-fashioned compared to microcircuits, biotechnology, and fluid organizations.

We now can see the outlines of a higher technology on a human scale, of a world with machines that don't clank, chemical plants that don't stink, and production systems that don't use people as cogs. Nanotechnology shows that advances can bring a different style of technology. Assemblers and AI will let us create complex products without complex organizations. Active shields will let us secure peace without a massive military-industrial complex. These technologies will broaden our choices by loosening our constraints, making room for greater diversity and independence. Establishing an era of universal wealth will require only that the vast, unclaimed resources of space someday be divided in a way that gives everyone a significant share.

In the next few sections, I will survey some extreme possibilities that new resources and new engines of creation will open for us—extremes that range from science-fiction to stone-age ways of life. Think of these extremes as intense primary colors, then mix your own palette to paint a future you like.

NANOTECHNOLOGY AND DAILY LIFE

Advancing technology may end or extend life, but it can also change its quality. Products based on nanotechnology will permeate the daily lives of people who choose to use them. Some consequences will be trivial; others may be profound.

Some products will have effects as ordinary as simplifying housekeeping (and as substantial as reducing the causes of domestic quarrels). It should be no great trick, for example, to make everything from dishes to carpets self-cleaning, and household air permanently fresh. For properly designed nanomachines, dirt would be food.

Other systems based on nanotechnology could produce fresh food —genuine meat, grain, vegetables, and so forth—in the home, year-round. These foods result from cells growing in certain patterns in plants and animals; cells can be coaxed to grow in these same patterns elsewhere. Home food growers will let people eat ordinary diets without killing anything. The animal rights movement (the forerunner of a movement to protect all conscious, feeling entities?) will be strengthened accordingly.

Nanotechnology will make possible high-resolution screens that project different images to each eye; the result will be three-dimensional television so real that the screen seems like a window into another world. Screens of this sort could line the helmet of a suit much like the spacesuit described in Chapter 6. The suit itself, rather than being programmed to transmit forces and textures from outside, could instead apply to the skin forces and textures defined by a complex, interactive program. A suit and helmet combination of this sort could simulate most of the sights and sensations of an entire environment, whether real or imaginary. Nanotechnology will make possible vivid art forms and fantasy worlds far more absorbing than any book, game, or movie.

Advanced technologies will make possible a whole world of prod-

ucts that make modern conveniences seem inconvenient and dangerous. Why shouldn't objects be light, flexible, durable, and cooperative? Why shouldn't walls look like whatever we want, and transmit
only the sounds we want to hear? And why should buildings and cars
ever crush or roast their occupants? For those who wish, the environment of daily life can resemble some of the wilder descriptions found
in science fiction.

OTHER SCIENCE FICTION DREAMS

Toward many extremes lie science fiction dreams, for those who
want to live them. They range from homes that cooperate with us for
our comfort to opportunities for toil on distant planets. Science fiction authors have imagined many things, some possible and others in
flat contradiction to known natural law. Some dreamed of spaceflight, and spaceflight came. Some dreamed of robots, and robots
came. Some dreamed of cheap spaceflight and intelligent robots, and
these too are coming. Other dreams seem possible.

Authors have written of the direct sharing of thoughts and emo-
tions from mind to mind. Nanotechnology seems likely to make possible some form of this by linking neural structures via transducers
and electromagnetic signals. Though limited to the speed of light,
this sort of telepathy seems as possible as telephony.

Starships, space settlements, and intelligent machines will all become possible. All this lies outside the skin, yet authors have written
also of transformations within the skin; these, too, will become possible. Becoming completely healthy in body and brain is one form of
change, yet some people will want more. They will seek changes on
a level deeper than mere health and wealth. Some will seek fulfillment in the world of the spirit; though that quest lies beyond the
scope of crude material technology, new physical possibilities will
provide new starting points and time enough to try. The technology
underlying cell repair systems will allow people to change their bodies in ways that range from the trivial to the amazing to the bizarre.
Such changes have few obvious limits. Some people may shed human
form as a caterpillar transforms itself to take to the air; others may
bring plain humanity to a new perfection. Some people will simply
cure their warts, ignore the new butterflies, and go fishing.

Authors have dreamed of time travel into the past, but nature
seems uncooperative. Yet biostasis opens travel into the future, since

it can make years pass in an eye blink. The jaded may seek the novelties of a more distant future, perhaps awaiting slowly maturing developments in the arts or society, or the mapping of the worlds of the galaxy. If so, they will choose sleep, passing from age to age in search of a time that suits them.

Strange futures lie open, holding worlds beyond our imagining.

ADVANCED SIMPLICITY

E. F. Schumacher, author of *Small Is Beautiful,* wrote: "I have no doubt that it is possible to give a new direction to technological development, a direction that shall lead it back to the real needs of man, and that also means: *to the actual size of man.* Man is small, and therefore small is beautiful." Schumacher was not writing of nanotechnology, but could such an advanced technology be part of a simpler life on a human scale?

In prehistoric times, people used two sorts of materials: the products of natural bulk processes (such as stone, water, air, and clay) and the products of natural molecular machinery (such as bone, wood, hide, and wool). Today we use these same materials and complex bulk processes to make the products of our global industrial civilization. If technological systems have grown past human scale, our bulk technology and stupid machines are largely to blame: to make systems complex, we have had to make them big. To make them capable, we have had to fill them with people. The resulting system now sprawls across continents, entangling people in a global web. It has offered escape from the toil of subsistence farming, lengthening lives and bringing wealth, but at a cost that some consider too high.

Nanotechnology will open new choices. Self-replicating systems will be able to provide food, health care, shelter, and other necessities. They will accomplish this without bureaucracies or large factories. Small, self-sufficient communities can reap the benefits.

One test of the freedom a technology offers is whether it frees people to return to primitive ways of life. Modern technology fails this test; molecular technology succeeds. As a test case, imagine returning to a stone-age style of life—not by simply ignoring molecular technology, but while using it.

Stone-age villagers lacking modern education wouldn't understand molecular machinery, but this matters little. Since ancient times, villagers have used the molecular machinery of yeast, seeds, and goats without molecular-level understanding. If such complex and unruly things as goats suit primitive ways of life, then other forms of molecular machinery will surely qualify. Living things show that the machinery inside a self-replicating system can be ignored in a way that the machinery inside an automobile cannot. Thus a group could raise novel "plants" and "animals" to ease the harsh edges of existence, and yet live a basically stone-age life. They could even limit themselves to ordinary plants and animals, engineered only by millennia of selective breeding.

With possibilities so broad, some people may even choose to live as we do today: with traffic noise, smells, and danger; with pitted teeth and whining drills; with aching joints and sagging skin; with joys offset by fear, toil, and approaching death. But unless they were brainwashed to obliterate their knowledge of better choices, how many people would willingly resign themselves to such lives? Perhaps a few.

Can one imagine living an ordinary life in a space settlement? A settlement would be large, complex, and located in space—but the Earth is also large, complex, and located in space. Worlds in space could be as self-maintaining as the Earth and as big as a continent, flooded with sunlight, filled with air, and holding a biocylinder if not a biosphere.

Worlds in space need not be products of direct human design. Underlying much of the beauty of nature is a certain kind of disorderly order. The veins on a leaf, the branches on a tree, the landforms in a watershed—all these have a freedom of form within patterns that resemble what mathematicians call "fractals." Lands in space need not be modeled on golf courses and suburban lots. Some will be shaped with the aid of computers programmed to reflect a deep knowledge of natural processes, melding human purpose with a natural quality that no human mind and hand can directly produce. Mountains and valleys in lands much like wilderness will mirror the shapes of dream-rock and dream-soil, sculpted by dream-ages of electronic water. Worlds in space will be worlds.

ROOM ENOUGH FOR DREAMS

This, then, is the size of the future's promise. Though limits to growth will remain, we will be able to harvest solar power a trillion times greater than all the power now put to human use. From the resources of our solar system, we will be able to create land area a million times that of Earth. With assemblers, automated engineering, and the resources of space we can rapidly gain wealth of a quantity and quality beyond past dreams. Ultimate limits to lifespan will remain, but cell repair technology will make perfect health and indefinitely long lives possible for everyone. These advances will bring new engines of destruction, but they will also make possible active shields and arms control systems able to stabilize peace.

In short, we have a chance at a future with room enough for many worlds and many choices, and with time enough to explore them. A tamed technology can stretch our limits, making the shape of technology pinch the shape of humanity less. In an open future of wealth, room, and diversity, groups will be free to form almost any sort of society they wish, free to fail or set a shining example for the world.

Unless your dreams demand that you dominate everyone else, chances are that other people will wish to share them. If so, then you and those others may choose to get together to shape a new world. If a promising start fails—if it solves too many problems or too few—then you will be able to try again. Our problem today is not to plan or build utopias but to seek a chance to try.

PREPARATIONS

We may fail. Replicating assemblers and AI will bring problems of unprecedented complexity, and they threaten to arrive with unprecedented abruptness. We cannot wait for a fatal error and *then* decide what to do about it; we must use these new technologies to build active shields before the threats are loosed.

Fortunately for our chances, the approaching breakthroughs will become steadily more obvious. They will eventually seize public attention, guaranteeing at least a measure of foresight. But the earlier we start planning, the better our chances. The world will soon become hospitable to memes that purport to describe sound policy for the assembler and AI breakthroughs. Such memes will then spread and become entrenched, whether they deserve to be or not. Our chances will be better if, when that time comes, a sound set of ideas has been hammered out and has begun to spread—public opinion and public policy will then be more likely to jump in a sensible direction when the crisis nears. This situation makes careful discussion and public education important right now. Guiding technology will also require new institutions, and institutions do not evolve overnight. This makes work on hypertext and fact forums important right now. If they are ready to use, then they too will grow more popular as the crisis nears.

Despite the broad appeal of an open future, some people will oppose it. The power-hungry, the intolerant idealists, and a handful of sheer people-haters will find the prospect of freedom and diversity repugnant. The question is, will they shape public policy? Governments will inevitably subsidize, delay, classify, manage, bungle, or guide the coming breakthroughs. The cooperating democracies may make a fatal error, but if they do, it will likely be the result of public confusion about which policies will have which consequences.

There will be genuine opposition to an open future, based on differing (and often unstated) values and goals, but there will be far greater disagreements over specific proposals, based on differing beliefs regarding matters of fact. And though many disagreements will stem from differences of judgment, many will inevitably stem from simple ignorance. Even solid, well-established facts will at first remain little known.

Worse, the prospect of technologies as fundamental as assemblers, AI, and cell repair machines must inevitably upset many old, entrenched ideas at once. This will cause conflicts in people's minds (I know; I've experienced some of them). In some minds, these conflicts will trigger the reject-the-new reflex that serves as humankind's most basic mental immune system. This reflex will make ignorance tenacious.

Worse yet, the spread of half-truths will also cause harm. To function properly, some memes must be linked to others. If the idea of nanotechnology were free from the idea of its danger, then nanotechnology would be a greater danger than it already is. But in a world grown wary of technology, this threat seems slight. Yet other idea fragments will spread, sowing misunderstanding and conflict.

The fact forum idea, when discussed without the distinctions among facts, values, and policies, will sound technocratic. Active shields, when proposed without mention of hypertext or fact forums, may seem impossible to trust. The danger and inevitability of nanotechnology, to those ignorant of active shields, will bring despair. The danger of nanotechnology, when its inevitability is not understood, will spur futile local efforts to stop its global advance. Active shields, when not motivated by the eventual requirements for controlling molecular technology, will strike some people as too much trouble. When called "defense projects," without distinction between defense and offense, shields will strike some people as threats to peace.

Likewise the idea of long life, when unaccompanied by the expectation of abundance and new frontiers, will seem perverse. Abundance, when imagined without space development or controlled replicators, will sound environmentally damaging. The idea of biostasis, to those who know nothing about cell repair and confuse expiration with dissolution, will sound absurd.

Unless they are held together by book covers or hypertext links, ideas will tend to split up as they travel. We need to develop and spread an understanding of the future as a whole, as a system of interlocking dangers and opportunities. This calls for the effort of many minds. The incentive to study and spread the needed information will be strong enough: the issues are fascinating and important, and many people will want their friends, families, and colleagues to join in considering what lies ahead. If we push in the right directions —learning, teaching, arguing, shifting directions, and pushing further—then we may yet steer the technology race toward a future with room enough for our dreams.

Eons of evolution and millennia of history have prepared this challenge and quietly presented it to our generation. The coming years will bring the greatest turning point in the history of life on Earth. To guide life and civilization through this transition is the great task of our time.

If we succeed (and if you survive) then you may be honored with endless questions from pesky great-grandchildren: "What was it like when you were a kid, back before the Breakthrough?" and "What was it like growing *old?*" and "What did you think when you heard the Breakthrough was coming?" and "What did you do then?" By your answers you will tell once more the tale of how the future was won.

AFTERWORD, 1990

WHAT WOULD I correct in *Engines* now, after several years of discussion, criticism, and technological progress? The first dozen pages would report recent advances in technology, but the conclusion would remain the same: we are moving toward assemblers, toward an era of molecular manufacturing giving thorough and inexpensive control of the structure of matter. There would be no changes in the central theses.

To summarize some indicators of technological progress: *Engines* speculates about when we might reach the milestone of designing a protein molecule from scratch, but this was actually accomplished in 1988 by William F. DeGrado of Du Pont and his colleagues. In 1987, a Nobel prize was shared by Donald J. Cram of UCLA, Jean-Marie Lehn of the Université Louis Pasteur, and Charles Pedersen of Du Pont for developing synthetic molecules with proteinlike capabilities. At IBM, John Foster's group has observed and modified individual molecules using the technology of the scanning tunneling microscope; this (or the related atomic force microscope) may within a few years provide a positioning mechanism for a crude protoassembler. Computer-based tools for designing and modeling molecules have improved rapidly. In short, advances toward nanotechnology through molecular systems engineering have been more rapid than *Engines* might suggest.

The idea of nanotechnology has spread far and wide, both through *Engines* itself (with 1990 editions in Japan and Britain) and through other publications. A recent summary appears in the 1990 Britannica yearbook, *Science and the Future*. I have been invited to speak at most of the top technical universities and many of the top corporate research laboratories in the United States. At Stanford, when I taught the first university course on nanotechnology, the room and hallway were packed on the first day, and the last entering student climbed through a window. Interest has been strong and growing.

What has been the reaction of the technical community—of those best placed to find and label erroneous ideas? From where I stand (e.g., in front of questioning technical audiences) the central theses of this book look solid; they have withstood criticism. This is not to say that everyone accepts them, merely that every reason suggested for rejecting them has turned out to be faulty. (My apologies to hidden critics with substantive points—please step out and speak up!) A variety of technical papers (on mechanical nanocomputers, molecular gears and bearings, etc.) are available and a technical book is on the way. After a series of local meetings, the Foresight Institute sponsored the first major conference on nanotechnology in October 1989 (covered in the November 4 *Science News*); a proceedings volume is in preparation.

At the conference, it became clear that Japan has for several years been treating molecular systems engineering as a basis for twenty-first-century technology. If the rest of the world wishes to see cooperative development of nanotechnology, it had best wake up and start doing its part.

Certain scenarios and proposals in the last third of *Engines* could bear rephrasing, but at least one problem is presented misleadingly. Page 173 speaks of the necessity of avoiding runaway accidents with replicating assemblers; today I would emphasize that there is little incentive to build a replicator even *resembling* one that can survive in nature. Consider cars: to work, they require gasoline, oil, brake fluid, and so forth. No mere accident could enable a car to forage in the wild and refuel from tree sap: this would demand engineering genius and hard work. It would be likewise with simple replicators designed to work in vats of assembler fluid, making nonreplicating products for use outside. Replicators built in accord with simple regulations would be unlike anything that could run wild. The problem—and it is enormous—is not one of accidents, but of abuse.

Some have mistakenly imagined that my aim is to promote nanotechnology; it is, instead, to promote understanding of nanotechnology *and its consequences*, which is another matter entirely. Nonetheless,

I am now persuaded that the sooner we start serious development efforts, the longer we will have for serious public debate. Why? Because serious debate will start with those serious efforts, and the sooner we start, the poorer our technology base will be. An early start will thus mean slower progress and hence more time to consider the consequences.

If you wish to keep up with developments in these areas, and with efforts to understand and influence them, please contact:

The Foresight Institute
P.O. Box 61058
Palo Alto, Calif. 94306
(415) 324-2490

NOTES AND REFERENCES

CHAPTER I: ENGINES OF CONSTRUCTION

P. 3 . . . Engines of Construction . . . The ideas in this chapter rest on technical arguments presented in my paper "Molecular Engineering: An Approach to the Development of General Capabilities for Molecular Manipulation" *(Proceedings of the National Academy of Sciences* (USA), Vol. 78, pp. 5275–78, 1981), which presents a case for the feasibility of designing protein molecules and developing general-purpose systems for directing molecular assembly.

P. 3 . . . "Protein engineering . . . represents . . ." See "Protein Engineering," by Kevin Ulmer *(Science,* Vol. 219, pp. 666–71, Feb. 11, 1983). Dr. Ulmer is now the director of the Center for Advanced Research in Biotechnology.

P. 5 . . . One dictionary . . . *The American Heritage Dictionary of the English Language,* edited by William Morris (Boston: Houghton Mifflin, 1978).

P. 6 . . . modern gene synthesis machines . . . See "Gene Machines: The Second Wave," by Jonathan B. Tucker *(High Technology,* pp. 50–59, March 1984).

P. 8 . . . other proteins serve basic mechanical functions . . . See Chapter 27 of *Biochemistry,* by Albert L. Lehninger (New York: Worth Publishers, 1975). This standard textbook is an excellent source of information on the molecular machinery of life. For a discussion of the bacterial flagellar motor, see "Ion Transport and the Rotation of Bacterial Flagella," by P. Lauger *(Nature,* Vol. 268, pp. 360–62, July 28, 1977).

P. 9 . . . self-assembling structures . . . For a description of molecular self-assembly, including that of the T4 phage and the ribosome, see Chapter 36 of Lehninger's *Biochemistry* (referenced above).

P. 9 . . . **Designing with Protein** . . . Nature has demonstrated a wide range of protein machines, but this will not limit us to designing with protein. For examples of fairly complex non-protein structures, see "Supramolecular Chemistry: Receptors, Catalysts, and Carriers," by Jean-Marie Lehn *(Science,* Vol. 227, pp. 849–56, February 22, 1985), which also speaks of designing "components, circuitry, and systems for signal and information treatment at the molecular level."

P. 9 . . . **any protein they can design** . . . Modern techniques can synthesize any desired DNA sequence, which can be used to direct ribosomes to make any desired amino acid sequence. Adding prosthetic groups is another matter, however.

P. 10 . . . **These tasks may sound similar** . . . For a comparison of the task of predicting natural protein structures with that of designing predictable structures, see "Molecular Engineering," referenced at the beginning of this section.

P. 10 . . . **in the journal** *Nature* . . . See "Molecular Technology: Designing Proteins and Peptides," by Carl Pabo *(Nature,* Vol. 301, p. 200, Jan. 20, 1983).

P. 10 . . . **short chains of a few dozen pieces** . . . See "Design, Synthesis, and Characterization of a 34-Residue Polypeptide That Interacts with Nucleic Acids," by B. Gutte et al. *(Nature,* Vol. 281, pp. 650–55, Oct. 25, 1979).

P. 10 . . . **They have designed from scratch a protein** . . . For a reference to this and a general discussion of protein engineering, see Kevin Ulmer's paper (referenced near the beginning of this section).

P. 10 . . . **changing their behaviors in predictable ways** . . . See "A Large Increase in Enzyme-Substrate Affinity by Protein Engineering," by Anthony J. Wilkinson et al. *(Nature,* Vol. 307, pp. 187–88, Jan. 12, 1984). Genetic engineering techniques have also been used to make an enzyme more stable, with no loss of activity. See "Disulphide Bond Engineered into T4 Lysozyme: Stabilization of the Protein Toward Thermal Inactivation," by L. Jeanne Perry and Ronald Wetzel of Genentech, Inc. *(Science,* Vol. 226, pp. 555–57, November 2, 1984).

P. 10 . . . **according to biologist Garrett Hardin** . . . in *Nature and Man's Fate* (New York: New American Library, 1959), p. 283.

P. 10 . . . **in the journal** *Science* . . . See "Biological Frontiers," by Frederick J. Blattner *(Science,* Vol. 222, pp. 719–20, Nov. 18, 1983).

P. 10 . . . **in** *Applied Biochemistry and Biotechnology* . . . See *Enzyme Engineering,* by William H. Rastetter *(Applied Biochemistry and Biotechnology,* Vol. 8, pp. 423–36, 1983). This review article describes several successful efforts to change the substrate specificity of enzymes.

P. 11 . . . two international workshops on molecular electronic devices . . . For the proceedings of the first, see *Molecular Electronic Devices*, edited by Forrest L. Carter (New York: Marcel Dekker, 1982). The proceedings of the second appear in *Molecular Electronic Devices II*, also edited by Forrest L. Carter (New York: Marcel Dekker, 1986). For a summary article, see "Molecular Level Fabrication Techniques and Molecular Electronic Devices," by Forrest L. Carter (*Journal of Vacuum Science and Technology*, B1(4), pp. 953–68, Oct.–Dec. 1983).

P. 11 . . . recommended support for basic research . . . See *The Institute* (a publication of the IEEE), January 1984, p. 1.

P. 11 . . . VLSI Research Inc. . . . Reported in *Microelectronic Manufacturing and Testing*, Sept. 1984, p. 49.

P. 12 . . . a single chemical bond . . . The strength of a single bond between two carbon atoms is about six nanonewtons, enough to support the weight of about 30,000 trillion carbon atoms. See *Strong Solids*, by A. Kelly, p. 12 (Oxford: Clarendon Press, 1973).

P. 12 . . . diamond fiber . . . Diamond is also over ten times stiffer than aluminum. See *Strong Solids* (referenced above), Appendix A, Table 2.

P. 13 . . . chemists . . . coax reacting molecules . . . See "Sculpting Horizons in Organic Chemistry," by Barry M. Trost *(Science, Vol. 227, pp. 908–16, February 22, 1985)*, which also mentions organic electrical conductors and the promise of molecular switches for molecular electronics.

P. 14 . . . will do all that proteins can do, and more . . . Chemists are already developing catalysts that improve on enzymes; see "Catalysts That Break Nature's Monopoly," by Thomas H. Maugh II *(Science, Vol. 221, pp. 351–54, July 22, 1983)*. For more on non-protein molecular tools, see "Artificial Enzymes," by Ronald Breslow *(Science, Vol. 218, pp. 532–37, November 5, 1982)*.

P. 14 . . . assemblers . . . See the first reference in this section. A device reported in 1982, called the scanning tunneling microscope, can position a sharp needle near a surface with an accuracy of a fraction of an atomic diameter. Besides demonstrating the feasibility of such positioning, it may be able to replace molecular machinery in positioning molecular tools. See "Scanning Tunneling Microscopy," by G. Binnig and H. Rohrer *(Physica 127B, pp. 37–45, 1985)*.

P. 14 . . . almost any reasonable arrangement . . . Assemblers will be able to create otherwise improbable arrangements of reactant molecules (overcoming entropy-of-activation factors), and will be able to direct the action of highly reactive chemical species. This will allow the use in controlled synthesis of reactions that would otherwise proceed only at a negligible rate or with an excessive number and rate of side reactions. Further, assemblers will be able to apply mechanical forces of bond-breaking magnitude to provide activation energy for reactions, and they will be able to employ molecular-scale conductors linked to voltage sources to manipulate electric fields in a direct and novel fashion. While photochemical techniques will not be as useful (because typical photon wavelengths are large on a molecular scale), similar results may sometimes be achieved by transfer of electronic excitation from molecule to molecule in a controlled, localized way.

Though assemblers will be powerful (and could even be directed to expand their own toolkits by assembling new tools), they will not be able to build everything that could exist. For example, a delicate structure might be designed that, like a stone arch, would self-destruct unless all its pieces were already in place. If there were no room in the design for the placement and removal of a scaffolding, then the structure might be impossible to build. Few structures of practical interest seem likely to exhibit such a problem, however. (In fact, the reversibility of the laws governing molecular motion implies that all *de*structable objects are, in principle, *con*structable; but if the destruction mechanisms all involve an explosive collapse, then attempts at construction via the reverse mechanism may have a negligible chance of success, owing to considerations involving the uncertainty of the trajectories of the incoming parts and the low entropy of the target state.)

P. 15 . . . the DNA-copying machinery in some cells . . . See "Comparative Rates of Spontaneous Mutation," by John W. Drake *(Nature,* Vol. 221, p. 1132, March 22, 1969). For a general discussion of this machinery, see Chapter 32 of Lehninger's *Biochemistry* (referenced above).

P. 15 . . . repairing and replacing radiation-damaged parts . . . The bacterium *Micrococcus radiodurans* has vigorous molecular repair mechanisms that enable it to survive the equivalent of more than a million years' worth of normal terrestrial background radiation delivered in a single dose. (See "Inhibition of Repair DNA Synthesis in *M. radiodurans* after Irradiation with Gamma-rays," by Shigeru Kitayama and Akira Matsuyama, in *Agriculture and Biological Chemistry,* Vol. 43, pp. 229–305, 1979.) This is about one thousand times the lethal radiation dose for humans, and enough to make Teflon weak and brittle.

P. 16 . . . life has never abandoned . . . Living organisms have built cell structures and simple molecular devices from lipids and sugars (and have built shells from silica and lime) but the lack of programmable assembly systems for these materials has kept life from exploiting them to form the main parts of complex molecular machines. RNA, like protein, has a structure directly determined by DNA, and it sometimes serves protein-like functions. See "First True RNA Catalyst Found" *(Science,* Vol. 223, p. 266, Jan. 20, 1984).

P. 17 . . . R. B. Merrifield . . . used chemical techniques . . . See Lehninger's *Biochemistry,* p. 119 (referenced above).

P. 18 . . . **during the mid-1800s, Charles Babbage** . . . See Chapter 2 of *Bit by Bit: An Illustrated History of Computers,* by Stan Augarten (New York: Ticknor & Fields, 1984).

P. 19 . . . **a billion bytes** . . . **in a box a micron wide** . . . If two different side groups on a polyethylene-like polymer are used to represent the ones and zeros of binary code, then the polymer can serve as a data-storage tape. If one were to use, say, fluorine and hydrogen as the two side groups, and to allow considerable room for tape reading, writing, and handling mechanisms, then a half cubic micron would store about a billion bytes. Access times can be kept in the microsecond range because the tapes can be made very short. A mechanical random-access memory scheme allows storage of only about 10 million bytes in the same volume, though this can probably be bettered. For a more detailed discussion, see "Molecular Machinery and Molecular Electronic Devices," by K. Eric Drexler, in *Molecular Electronic Devices II,* edited by Forrest L. Carter (New York: Marcel Dekker, 1986).

P. 19 . . . **mechanical signals** . . . These could be sent by pushing and pulling atom-wide rods of carbyne, a form of carbon in which the atoms are linked in a straight line by alternating single and triple bonds. See "Molecular Machinery and Molecular Electronic Devices," referenced in the above note.

P. 19 . . . **a scheme proposed by** . . . **Richard Feynman** . . . See his article "Quantum Mechanical Computers" *(Optics News,* Vol. 11, pp. 11–20, Feb. 1985). Feynman concludes that "the laws of physics present no barrier to reducing the size of computers until bits are the size of atoms, and quantum behavior holds dominant sway."

P. 19 . . . **a** *disassembler* . . . There will be limits to disassemblers as well: For example, one could presumably design a sensitive structure that would fall apart (or explode) when tampered with, preventing controlled disassembly.

CHAPTER 2: THE PRINCIPLES OF CHANGE

P. 21 . . . **"Think of the design process** . . ." See *The Sciences of the Artificial* (Second Edition) by Herbert A. Simon (Cambridge, Mass.: MIT Press, 1981). This book explores a range of issues related to engineering, problem-solving, economics, and artificial intelligence.

P. 23 . . . **Both strand and copy** . . . Because of the rules for nucleotide pairing, the copies actually resemble photographic negatives, and only a copy of a copy matches the original itself.

P. 23 . . . **Biochemist Sol Spiegelman** . . . A discussion of his work in this area appears in "The Origin of Genetic Information," by Manfred Eigen et al. *(Scientific American,* Vol. 244, pp. 88–117, April 1981).

P. 23 . . . **Oxford zoologist Richard Dawkins** . . . discusses replicators in *The Selfish Gene* (New York: Oxford University Press, 1976). This readable book offers an excellent introduction to modern concepts of evolution, focusing on germ-line replicators as the units that undergo variation and selection in evolution.

P. 26 . . . As Richard Dawkins points out . . . in *The Selfish Gene* (see above).

P. 27 . . . Darwin's detested book . . . *The Origin of Species*, by Charles R. Darwin (London: Charles Murray, 1859).

P. 28 . . . the . . . ideas of evolution were known before Darwin . . . See p. 59 of *The Constitution of Liberty*, by Friedrich A. Hayek (Chicago: University of Chicago Press, 1960) for a discussion of the earlier work on linguistic, institutional, and even biological evolution, which apparently developed "the conceptual apparatus that Darwin employed." See also p. 23 of *Law, Legislation and Liberty*, Vol. 1, *Rules and Order* (Chicago: University of Chicago Press, 1973). Elsewhere, these books discuss the concept of liberty under law and the crucial distinction between a law and a command. These will be important to matters discussed in Chapters 11 and 12.

P. 29 . . . As Richard Dawkins puts it . . . in *The Selfish Gene* (see above).

P. 31 . . . in *The Next Whole Earth Catalog* . . . Edited by Stewart Brand (Sausalito, California: POINT; distributed by Random House, New York, 1980).

P. 31 . . . Peters and Waterman . . . See *In Search of Excellence: Lessons from America's Best-Run Corporations*, by Thomas J. Peters and Robert H. Waterman, Jr. (New York: Warner Books, 1982).

P. 33 . . . as Alfred North Whitehead stated . . . in *Science and the Modern World* (New York: Macmillan Company, 1925).

P. 34 . . . only study, imagination, and thought . . . Converting these into good computer graphics and video will help a lot, though.

P. 35 . . . Richard Dawkins calls . . . "Meme" is a meme that was launched in the last chapter of *The Selfish Gene* (see above).

P. 36 . . . selfish motives can encourage cooperation . . . In *The Evolution of Cooperation* (New York: Basic Books, 1984) political scientist Robert Axelrod uses a multisided computer game and historical examples to explore the conditions required for cooperation to evolve among selfish entities. Being nice, retaliatory, *and* forgiving is important to evolving stable cooperation. Chapter 7 of this valuable book discusses "How to Promote Cooperation."

P. 36 . . . In *The Extended Phenotype* . . . by Richard Dawkins (San Francisco: W. H. Freeman, 1982).

P. 37 . . . This meme package infected the Xhosa people . . . See

"The Self-Destruction of the Xosas," Elias Canetti, *Crowds and Power* (New York: Continuum, 1973), p. 193.

CHAPTER 3: PREDICTING AND PROJECTING

P. 39 . . . "The critical attitude . . ." From *Conjectures and Refutations: The Growth of Scientific Knowledge*, by Sir Karl Popper (New York: Basic Books, 1962).

P. 40 . . . Richard Feynman . . . gave a talk . . . "There's Plenty of Room at the Bottom," reprinted in *Miniaturization*, edited by H. D. Gilbert (New York: Reinhold, 1961).

P. 42 . . . Bertrand Russell observed . . . Quoted by Karl Popper in *Objective Knowledge: An Evolutionary Approach* (Oxford: Clarendon Press, 1972).

P. 43 . . . to *seem* true or . . . to *be* true . . . Ideas that have evolved to *seem* true (at least to uncritical minds) can in fact be quite false. An excellent work that compares naïve human judgment to judgment aided by scientific and statistical techniques is *Human Inference*, a book by Richard Nisbett and Lee Ross in the Century Psychology Series (Englewood Cliffs, New Jersey: Prentice-Hall, 1980). It shows that, just as we suffer from optical illusions and blind spots, so we suffer from cognitive illusions and blind spots. Other experiments show that untutored people share systematic misunderstandings of such elementary facts as the direction a ball will move when whirled in a circle and then released; learned medieval philosophers (who neglected to test their ideas against reality) evolved whole systems of "science" based on identical misunderstandings. See "Intuitive Physics," by Michael McClosky *(Scientific American,* Vol. 248, pp. 122–30, Apr. 1983).

P. 44 . . . survivors . . . huddle so close together . . . Strictly speaking, this applies only to survivors that are themselves uniform, general theories. The theory that all rocks will fall straight up next Wednesday has not been disproved (and would have practical consequences), but the special reference to Wednesday makes it nonuniform.

P. 44 . . . as Karl Popper points out . . . See his *Logic of Scientific Discovery,* pp. 124 and 419 (New York: Harper & Row, 1965). See also *Objective Knowledge,* p.15.

P. 45 . . . As . . . Ralph E. Gomory says . . . in "Technology Development" *(Science,* Vol. 220, pp. 576–80, May 6, 1983).

P. 48 . . . their interplay of function and motion . . . These were clear only at low speeds; while Leonardo surely had some intuitive sense of dynamics, a description of dynamics adequate to predict the behavior of fast-moving, high-acceleration machine parts did not arrive until Newton.

P. 49 . . . designs even in the absence of the tools . . . Familiarity with the steady progress in chip fabrication technology has led some companies to design microprocessors whose manufacture required techniques not available at the time of their design.

P. 49 . . . computer-aided design of molecular systems . . . To do this well will require the simulation of molecular systems. A discussion of one system for molecular simulation appears in Robert Bruccoleri's doctoral thesis, "Macromolecular Mechanics and Protein Folding" (Harvard University, May 1984). For the results of a simulation, see "Dynamics and Conformational Energetics of a Peptide Hormone: Vasopressin," by A. T. Hagler et al. *(Science,* Vol. 227, pp. 1309–15, Mar. 15, 1985). These references both describe *classical* simulations, which describe how molecules move in response to forces; such simulations will be adequate for most parts of a typical molecular machine. Other work requires more fundamental (and more costly) *quantum mechanical* simulations, which describe the distribution of electrons in molecules. These calculations will be required to describe the forming and breaking of bonds by assembler tools. For a discussion of molecular simulations that include quantum mechanical calculations of bond formation, see "Theoretical Chemistry Comes Alive: Full Partner with Experiment," by William H. Goddard III *(Science,* Vol. 227, pp. 912–23, Feb. 22, 1985). See also *Lecture Notes in Chemistry, 19, Computational Aspects for Large Chemical Systems,* by Enrico Clementi (New York: Springer-Verlag, 1980). Finally, for a discussion of present design tools, see "Designing Molecules by Computer," by Jonathan B. Tucker *(High Technology,* pp. 52–59, Jan. 1984). Parallel-processing computers will greatly aid computational chemistry and computer-aided design.

P. 50 . . . design ahead . . . Early design-ahead efforts seem likely to aim at defining a workable assembler system; it need not be ideal, so long as it has fairly broad capabilities. Once the capabilities of this standard assembler design are fairly well specified—even before the design is complete—it will become possible (1) to begin developing a library of nanomachine designs suited to construction by this standard assembler (or by assemblers that the standard assembler could construct), and (2) to prepare a corresponding library of procedures for the assembly of these designs. Then when the first crude assembler is developed, it can be used (perhaps through an intermediate stage of tool building) to build a standard assembler. This in turn could be used to build anything in the design library.

Early assemblers will greatly extend our ability to make things. With even limited design ahead, the advent of assemblers will result almost immediately in substantial jumps in the quality of hardware. Since assemblers will be built by assemblers, some form of self-replicating system will be an immediate natural consequence of design ahead and the assembler breakthrough. Accordingly, the advent of assemblers may make possible not only a jump in hardware quality, but the almost immediate mass production of that hardware in unprecedented quantities (see Chapter 4). For better or for worse, this will make possible an unusually abrupt change in technology, in economics, and in world affairs.

CHAPTER 4: ENGINES OF ABUNDANCE

P. 53 . . . If every tool, when ordered . . . From *Scientific Quotations: The Harvest of a Quiet Eye*, selected by A. L. Mackay, edited by M. Ebison (New York: Crane, Russak, 1977).

P. 53 . . . a NASA scientist . . . Former NASA administrator Robert Frosch said much the same thing at the IEEE Centennial Technical Convocation (see *The Institute*, p. 6, Dec. 1984).

P. 54 . . . replicators, such as viruses, bacteria . . . In an evolutionary sense, an animal's genes are replicators, but the animal itself is not; only changes to genes, not changes to an animal's body, are replicated in later generations. This distinction between genetic replicators and the systems they shape is essential to understanding evolution, but use of the term "replicator" to refer to the whole system is more convenient when discussing replicating systems as productive assets.

P. 54 . . . Fujitsu Fanuc . . . See "Production: A Dynamic Challenge," by M. E. Merchant *(IEEE Spectrum,* pp. 36–39, May 1983). This issue of the *IEEE Spectrum* contains an extensive discussion of computer-based automation.

P. 56 . . . will instead resemble factories . . . Cell-style organization nonetheless has advantages. For example, despite various active-transport mechanisms, cells typically transport molecular components by diffusion rather than by conveyors. This effectively connects every machine to every other (in the same membrane compartment) in a robust fashion; conveyors, in contrast, can break down, requiring repair or replacement. But it may be that properly implemented conveyor-based transportation has strong advantages and yet did not evolve. Conveyor-based systems would be harder to evolve because they require a new molecular machine to have a suitable location, orientation, and interface to the conveyor before it can function. If it failed to meet any of these requirements it would be useless, and selective pressures would generally eliminate it before a useful variant had a chance to appear. For a new molecular machine to function in a diffusion-based system, though, it need only be present. If it does something useful, selection will favor it immediately.

P. 57 . . . A fast enzyme . . . See Albert L. Lehninger's *Biochemistry,* p. 208 (in Chapter 1 references). Further, each molecule of the enzyme catalase can break down 40 million hydrogen peroxide molecules per second; see *Enzyme Structure and Mechanism,* by Alan Fersht, p. 132 (San Francisco: W. H. Freeman & Co., 1977). In typical enzymatic reactions, molecules must wander into position with respect to the enzyme's "tools," then wait for random thermal vibrations to cause a reaction, and then wander out of the way again. These steps take up almost all of the enzyme's time; the time required to form or break a bond is vastly smaller. Because the electrons of a bond are over a thousand times lighter and more mobile than the nuclei that define atomic positions, the slower motion of whole atoms sets the pace. The speed of typical atoms under thermal agitation at ordinary temperatures is over 100 meters per second, and the distance an atom must move to form or break a bond is typically about a ten billionth of a meter, so the time required is about one trillionth of a second. See Chapter 12 of *Molecular Thermodynamics*, by John H. Knox (New York: Wiley-Interscience, 1971).

P. 57 . . . about fifty million times more rapidly . . . This scaling relationship may be verified by observing (1) that mechanical disturbances travel at the speed of sound (arriving in half the time if they travel half as far) and (2) that, for a constant stress in the arm material, reducing the arm length (and hence the mass per unit cross section) by one half doubles the acceleration at the tip while halving the distance the tip travels, which allows the tip to move back and forth in half the time (since the time required for a motion is the square root of a quantity proportional to the distance traveled divided by the acceleration).

P. 57 . . . a copy . . . of the tape . . . Depending on the cleverness (or lack of cleverness) of the coding scheme, the tape might have more mass than the rest of the system put together. But since tape duplication is a simple, specialized function, it need not be performed by the assembler itself.

P. 59 . . . only a minute fraction misplaced . . . due to rare fluctuations in thermal noise, and to radiation damage during the assembly process. High-reliability assemblers will include a quality-control system to identify unwanted variations in structure. This system could consist of a sensor arm used to probe the surface of the workpiece to identify the unplanned bumps or hollows that would mark a recent mistake. Omissions (typically shown by hollows) could be corrected by adding the omitted atoms. Misplaced groups (typically shown by bumps) could be corrected by fitting the assembler arm with tools to remove the misplaced atoms. Alternatively, a small workpiece could simply be completed and tested. Mistakes could then be discarded before they had a chance to be incorporated into a larger, more valuable system. These quality-control steps will slow the assembly process somewhat.

P. 62 . . . working, like muscle . . . See the notes for Chapter 6.

CHAPTER 5: THINKING MACHINES
P. 64 . . . The world stands . . . Quoted from *Business Week*, March 8, 1982.

P. 65 . . . As Daniel Dennett . . . points out . . . See "Why the Law of Effect Will Not Go Away," in Daniel C. Dennett, *Brainstorms: Philosophical Essays on Mind and Psychology* (Cambridge, Mass.: MIT Press, 1981). This book explores a range of interesting issues, including evolution and artificial intelligence.

P. 66 . . . Marvin Minsky . . . views the mind . . . See his book *The Society of Mind* (New York: Simon & Schuster, to be published in 1986). I have had an opportunity to review much of this work in manuscript form; it offers valuable insights about thought, language, memory, developmental psychology, and consciousness—and about how these relate to one another and to artificial intelligence.

P. 67 . . . Any system or device . . . *American Heritage Dictionary,* edited by William Morris (Boston: Houghton Mifflin Company, 1978).

P. 68 . . . Babbage had built . . . See *Bit by Bit,* in Chapter 1 references.

P. 68 . . . the *Handbook of Artificial Intelligence* . . . edited by Avron Barr and Edward A. Feigenbaum (Los Altos, Calif.: W. Kaufmann, 1982).

P. 69 . . . As Douglas Hofstadter urges . . . See "The Turing Test: A Coffeehouse Conversation" in *The Mind's I,* composed and arranged by Douglas R. Hofstadter and Daniel C. Dennett (New York: Basic Books, 1981).

P. 69 . . . We can surely make machines . . . As software engineer Mark Miller puts it, "Why should people be able to make intelligence in the bedroom, but not in the laboratory?"

P. 70 . . . "I believe that by the end of the century . . ." From "Computing Machinery and Intelligence," by Alan M. Turing *(Mind,* Vol. 59, No. 236, 1950); excerpted in *The Mind's I* (referenced above).

P. 71 . . . a system could show both kinds . . . Social and technical capabilities might stem from a common basis, or from linked subsystems; the boundaries can easily blur. Still, specific AI systems could be clearly deserving of one name or the other. Efforts to make technical AI systems as useful as possible will inevitably involve efforts to make them understand human speech and desires.

P. 71 . . . *social AI* . . . Advanced social AI systems present obvious dangers. A system able to pass the Turing test would have to be able to plan and set goals as a human would—that is, it would have to be able to plot and scheme, perhaps to persuade people to give it yet more information and ability. Intelligent people have done great harm through words alone, and a Turing-test passer would of necessity be designed to understand and deceive people (and it would not necessarily have to be inbued with rigid ethical standards, though it might be). Chapter 11 discusses the problem of how to live with advanced AI systems, and how to build AI systems worthy of trust.

P. 71 . . . "May not machines carry out . . ." See the earlier reference to Turing's paper, "Computing Machinery and Intelligence."

P. 72 . . . Developed by Professor Douglas Lenat . . . and described by him in a series of articles on "The Nature of Heuristics" (*Artificial Intelligence,* Vol. 19, pp. 189–249, 1982; Vol. 21, pp. 31–59 and 61–98, 1983; Vol. 23, pp. 269–93, 1984).

P. 73 . . . Traveller TCS . . . See preceding reference, Vol. 21, pp. 73–83.

P. 74 . . . EURISKO has shortcomings . . . Lenat considers the most serious to be EURISKO's limited ability to evolve new *representations* for new information.

P. 75 . . . In October of 1981 . . . In the fall of 1984 . . . See "The 'Star Wars' Defense Won't Compute," by Jonathan Jacky (*The Atlantic,* Vol. 255, pp. 18–30, June 1985).

P. 75 . . . in the *IEEE Spectrum* . . . See "Designing the Next Generation," by Paul Wallich (*IEEE Spectrum,* pp. 73–77, November 1983).

P. 76 . . . fresh insights into human psychology . . . Hubert Dreyfus, in his well-known book *What Computers Can't Do: The Limits of Artificial Intelligence* (New York: Harper & Row, 1979), presents a loosely reasoned philosophical argument that digital computers can *never* be programmed to perform the full range of human intellectual activities. Even if one were to accept his arguments, this would not affect the main conclusions I draw regarding the future of AI: the automation of engineering design is not subject to his arguments because it does not require what he considers genuine intelligence; duplicating the human mind by means of neural simulation avoids (and undermines) his philosophical arguments by dealing with mental processes at a level where those arguments do not apply.

P. 77 . . . virus-sized molecular machines . . . See Chapter 7.

P. 77 . . . build analogous devices . . . These devices might be electromechanical, and will probably be controlled by microprocessors; they will not be as simple as transistors. Fast neural simulation of the sort I describe will be possible even if each simulated synapse must have its properties controlled by a device as complex as a microprocessor.

P. 78 . . . experimental electronic switches . . . which switch in slightly over 12 picoseconds are described in "The HEMT: A Superfast Transistor," by Hadis Morkoc and Paul M. Solomon (*IEEE Spectrum,* pp. 28–35, Feb. 1984).

P. 78 . . . Professor Robert Jastrow . . . in his book *The Enchanted Loom: The Mind in the Universe* (New York: Simon & Schuster, 1981).

P. 79 . . . will fit in less than a cubic centimeter . . . The brain consists chiefly of wirelike structures (the axons and dendrites) and switchlike structures (the synapses). This is an oversimplification, however, because at least some wirelike structures can have their resistance modulated on a short time scale (as discussed in "A Theoretical Analysis of Electrical Properties of Spines," by C. Koch and T. Poggio, MIT AI Lab Memo No. 713, April 1983). Further, synapses behave less like switches than like modifiable switching circuits; they can be modulated on a short time scale and entirely rebuilt on a longer time scale (see "Cell Biology of Synaptic Plasticity," by Carl W. Cotman and Manuel Nieto-Sampedro, *Science*, Vol. 225, pp. 1287–94, Sept. 21, 1984).

The brain can apparently be modeled by a system of nanoelectronic components modulated and rebuilt by nanomachinery directed by mechanical nanocomputers. Assume that one nanocomputer is allotted to regulate each of the quadrillion or so "synapses" in the model brain, and that each also regulates corresponding sections of "axon" and "dendrite." Since the volume of each nanocomputer (if equivalent to a modern microprocessor) will be about 0.0003 cubic micron (See "Molecular Machinery and Molecular Electronic Devices," referenced in Chapter 1), these devices will occupy a total of about 0.3 cubic centimeter. Dividing another 0.3 cubic centimeter equally between fast random-access memory and fairly fast tape memory would give each processor a total of about 3.7K bytes of RAM and 275K bytes of tape. (This sets no limit to program complexity, since several processors could share a larger program memory.) This amount of information seems *far* more than enough to provide an adequate model of the functional state of a synapse. Molecular machines (able to modulate nanoelectronic components) and assembler systems (able to rebuild them) would occupy comparatively little room. Interchange of information among the computers using carbyne rods could provide for the simulation of slower, chemical signaling in the brain.

Of the nanoelectronic components, wires will occupy the most volume. Typical dendrites are over a micron in diameter, and serve primarily as conductors. The diameter of thin wires could be less than a hundredth of a micron, determined by the thickness of the insulation required to limit electron tunneling (about three nanometers at most). Their conductivity can easily exceed that of a dendrite. Since the volume of the entire brain is about equal to that of a ten-centimeter box, wires a hundred times thinner (one ten-thousandth the cross section) will occupy at most 0.01 of a cubic centimeter (allowing for their being shorter as well). Electromechanical switches modulated by molecular machinery can apparently be scaled down by about the same factor, compared to synapses.

Thus, nanoelectronic circuits that simulate the electrochemical behavior of the brain can apparently fit in a bit more than 0.01 cubic centimeter. A generous allowance of volume for nanocomputers to simulate the slower functions of the brain totals 0.6 cubic centimeter, as calculated above. A cubic centimeter thus seems ample.

P. 79 . . . dissipating a millionfold more heat . . . This may be a pessimistic assumption, however. For example, consider axons and dendrites as electrical systems transmitting signals. All else being equal, millionfold faster operation requires millionfold greater currents to reach a given voltage threshold. Resistive heating varies as the current squared, divided by the conductivity. But copper has about *forty* million times the conductivity of neurons (see "A Theoretical Analysis of Electrical Properties of Spines," referenced above), reducing resistive heating to less than the level assumed (even in a device like that described in the text, which is somewhat more compact than a brain). For another example, consider the energy dissipated in the triggering of a synapse: devices requiring less energy per triggering would result in a power dissipation less than that assumed in the text. There seems no reason to believe that neurons are near the limits of energy efficiency in information processing; for a discussion of where those limits lie, see "Thermodynamics of Computation—A Review," by C. H. Bennett *(International Journal of Theoretical Physics,* Vol. 21, pp. 219–53, 1982). This reference states that neurons dissipate an energy of over one billion electron-volts per discharge. Calculations indicate that electrostatically activated mechanical relays can switch on or off in less than a nanosecond, while operating at less than 0.1 volt (like neurons) and consuming less than a hundred electron-volts per operation. (There is no reason to believe that mechanical relays will make the best switches, but their performance is easy to calculate.) Interconnect capacitances can also be far lower than those in the brain.

P. 79 . . . pipe . . . bolted to its top . . . This is a somewhat silly image, since assemblers can make connectors that work better than bolts and cooling systems that work better than flowing water. But to attempt to discuss systems based entirely on advanced assembler-built hardware would at best drag in details of secondary importance, and would at worst sound like a bogus prediction of what *will* be built, rather than a sound projection of what *could* be built. Accordingly, I will often describe assembler-built systems in contexts that nanotechnology would in fact render obsolete.

P. 79 . . . As John McCarthy . . . points out . . . See *Machines Who Think,* by Pamela McCorduck, p. 344 (San Francisco: W. H. Freeman & Company, 1979). This book is a readable and entertaining overview of artificial intelligence from the perspective of the people and history of the field.

P. 81 . . . As Marvin Minsky has said . . . in *U.S. News & World Report,* p. 65, November 2, 1981.

CHAPTER 6: THE WORLD BEYOND EARTH

P. 86 . . . engineers know of alternatives . . . For orbit-to-orbit transportation, one attractive alternative is using rockets burning fuel produced in space from space resources.

P. 86 . . . result will be the "lightsail" . . . For further discussion, see "Sailing on Sunlight May Give Space Travel a Second Wind" *(Smithsonian,* pp. 52–61, Feb. 1982), "High Performance Solar Sails and Related Reflecting Devices," AIAA Paper 79-1418, in *Space Manufacturing III,* edited by Jerry Grey and Christine Krop (New York: American Institute of Aeronautics and Astronautics, 1979), and MIT Space Systems Laboratory Report 5-79, by K. Eric Drexler. The World Space Foundation (P.O. Box Y, South Pasadena, Calif. 91030) is a nonprofit, membership-oriented organization that is building an experimental solar sail and supporting the search for accessible asteroids.

P. 88 . . . The asteroids . . . are flying mountains of resources . . . For a discussion of asteroidal resources, see "Asteroid Surface Materials: Mineralogical Characterizations from Reflectance Spectra," by Michael J. Gaffey and Thomas B. McCord *(Space Science Reviews,* No. 21, p. 555, 1978) and "Finding 'Paydirt' on the Moon and Asteroids," by Robert L. Staehle *(Astronautics and Aeronautics,* pp. 44–49, November 1983).

P. 88 . . . as permanent as a hydroelectric dam . . . Erosion by micrometeoroids is a minor problem, and damage by large meteoroids is extremely rare.

P. 88 . . . Gerard O'Neill . . . See his book *The High Frontier: Human Colonies in Space* (New York: William Morrow, 1976). The Space Studies Institute (285 Rosedale Road, P.O. Box 82, Princeton, N.J. 08540) is a nonprofit, membership-oriented organization aimed at advancing the economic development and settlement of space, working chiefly through research projects. The L5 Society (1060 East Elm, Tucson, Ariz. 85719) is a nonprofit, membership-oriented organization aimed at advancing the economic development and settlement of space, working chiefly through public education and political action.

P. 90 . . . Using this energy to power assemblers . . . How much electric power can a given mass of solar collector supply? Since electric energy is readily convertible to chemical energy, this will indicate how rapidly a solar collector of a given mass can supply enough energy to construct an equal mass of something else. Experimental amorphous-silicon solar cells convert sunlight to electricity with about 10 percent efficiency in an active layer about a micron thick, yielding about 60 kilowatts of power per kilogram of active mass. Assembler-built solar cells will apparently be able to do much better, and need not have heavy substrates or heavy, low-voltage electrical connections. Sixty kilowatts of power supplies enough energy in a few minutes to break and rearrange all the chemical bonds in a kilogram of typical material. Thus a spacecraft with a small fraction of its mass invested in solar collectors will be able to entirely rework its own structure in an hour or so. More important, though, this calculation indicates that solar-powered replicators will be able to gather enough power to support several doublings per hour.

P. 91 . . . The middle layer of the suit material . . . To have the specified strength, only about one percent of the material's cross-sectional area must consist of diamond fibers (hollow telescoping rods, in one implementation) that run in a load-bearing direction. There exists a regular, three-dimensional woven pattern (with fibers running in seven different directions to support all possible types of load, including shear) in which packed cylindrical fibers fill about 45 percent of the total volume. In any given direction, only some of the fibers can bear a substantial load, and using hollow, telescoping fibers (and then extending them by a factor of two in length) makes the weave less dense. These factors consume most of the margin between 45 percent and one percent, leaving the material only as strong as a typical steel.

For the suit to change shape while holding a constant internal volume at a constant pressure, and do so efficiently, the mechanical energy absorbed by stretching material in one place must be recovered and used to do mechanical work in contracting material in another place—say, on the other side of a bending elbow joint. One way to accomplish this is by means of electrostatic motors, reversible as generators, linked to a common electric power system. Scaling laws favor electrostatic over electromagnetic motors at small sizes.

A design exercise (with applications not limited to hypothetical space suits) resulted in a device about 50 nanometers in diameter that works on the principle of a "pelletron"-style Van de Graaff generator, using electron tunneling across small gaps to charge pellets and using a rotor in place of a pellet chain. (The device also resembles a bucket-style water wheel.) DC operation would be at 10 volts, and the efficiency of power conversion (both to and from mechanical power) seems likely to prove excellent, limited chiefly by frictional losses. The power-conversion density (for a rotor rim speed of one meter per second and pellets charged by a single electron) is about three trillion watts per cubic meter. This seems more than adequate.

As for frictional losses in general, rotary bearings with strengths of over 6 nanonewtons can be made from carbon bonds—see *Strong Solids*, by A. Kelly (Oxford: Clarendon Press, 1973)—and bearings using a pair of triple-bonded carbon atoms should allow almost perfectly unhindered rotation. Roller bearings based on atomically perfect hollow cylinders with bumps rolling gear-fashion on atomically perfect races have at least two significant energy-dissipation modes, one resulting from phonon (sound) radiation through the slight bumpiness of the rolling motion, the other resulting from scattering of existing phonons by the moving contact point. Estimates of both forms of friction (for rollers at least a few nanometers in diameter moving at modest speeds) suggest that they will dissipate very little power, by conventional standards.

Electrostatic motors and roller bearings can be combined to make telescoping jackscrews on a submicron scale. These can in turn be used as fibers in a material able to behave in the manner described in the text.

P. 91 . . . now transmits only a tenth of the force . . . An exception to this is a force that causes overall acceleration: for example, equilibrium demands that the forces on the soles of the feet of a person standing in an accelerating rocket provide support, and the suit must transmit them without amplification or diminution. Handling this smoothly may be left as an exercise for future control-system designers and nanocomputer programmers.

P. 92 . . . the suit will keep you comfortable . . . Disassemblers, assemblers, power, and cooling—together, these suffice to recycle all the materials a person needs and to maintain a comfortable environment. Power and cooling are crucial.

As for power, a typical person consumes less than 100 watts, on the average; the solar power falling on a surface the size of a sheet of typing paper (at Earth's distance from the Sun) is almost as great. If the suit is covered with a film that acts as a high-efficiency solar cell, the sunlight striking it should provide enough power. Where this is inadequate, a solar-cell parasol could be used to gather more power.

As for cooling, all power absorbed must eventually be disposed of as waste heat—in a vacuum, by means of thermal radiation. At body temperature, a surface can radiate over 500 watts per square meter. With efficient solar cells and suitable design (and keeping in mind the possibility of cooling fins and refrigeration cycles), cooling should be no problem in a wide range of environments. The suit's material can, of course, contain channels for the flow of coolant to keep the wearer's skin at a preferred temperature.

P. 92 . . . a range of devices greater than . . . yet built . . . A pinhead holds about a cubic millimeter of material (depending on the pin, of course). This is enough room to encode an amount of text greater than that in a *trillion* books (large libraries hold only millions). Even allowing for a picture's being worth a thousand words, this is presumably enough room to store plans for a wide enough range of devices.

P. 92 . . . in a morning . . . The engineering AI systems described in Chapter 5, being a million times faster than human engineers, could perform several centuries' worth of design work in a morning.

P. 92 . . . replicating assemblers that work in space . . . Assemblers in a vacuum can provide any desired environment at a chemical reaction site by positioning the proper set of molecular tools. With proper design and active repair-and-replacement mechanisms, exposure to the natural radiation of space will be no problem.

P. 95 . . . move it off Earth entirely . . . But what about polluting space? Debris in Earth orbit is a significant hazard and needs to be controlled, but many environmental problems on Earth cannot occur in space: space lacks air to pollute, groundwater to contaminate, or a biosphere to damage. Space is already flooded with natural radiation. As life moves into space, it will be protected from the raw space environment. Further, space is *big*—the volume of the inner solar system alone is many trillions of times that of Earth's air and oceans. If technology on Earth has been like a bull in a china shop, then technology in space will be like a bull in an open field.

P. 96 . . . As Konstantin Tsiolkovsky wrote . . . Quoted in *The High Frontier: Human Colonies in Space,* by Gerard K. O'Neill (New York: William Morrow, 1976).

P. 96 . . . drive a beam far beyond our solar system . . . This concept was first presented by Robert L. Forward in 1962.

P. 96 . . . Freeman Dyson . . . suggests . . . He discussed this in a talk at an informal session of the May 15, 1980, "Discussion Meeting on Gossamer Spacecraft," held at the Jet Propulsion Laboratory in Pasadena, California.

P. 96 . . . Robert Forward . . , suggests . . . See his article "Round-trip Interstellar Travel Using Laser-Pushed Lightsails," (*Journal of Spacecraft and Rockets,* Vol. 21, pp. 187–95, Jan.–Feb. 1984). Forward notes the problem of making a beam-reversal sail light enough, yet of sufficient optical quality (diffraction limited) to do its job. An actively controlled structure based on thin metal films positioned by nanometer-scale actuators and computers seems a workable approach to solving this problem.

But nanotechnology will allow a different approach to accelerating lightsails and stopping their cargo. Replicating assemblers will make it easy to build large lasers, lenses, and sails. Sails can be made of a crystalline dielectric, such as aluminum oxide, having extremely high strength and low optical absorptivity. Such sails could endure intense laser light, scattering it and accelerating at many gees, approaching the speed of light in a fraction of a year. This will allow sails to reach their destinations in near-minimal time. (For a discussion of the multi-gee acceleration of dielectric objects, see "Applications of Laser Radiation Pressure," by A. Ashkin [*Science,* Vol. 210, pp. 1081–88, Dec. 5, 1980].)

In flight, computer-driven assembler systems aboard the sail (powered by yet more laser light from the point of departure) could rebuild the sail into a long, thin traveling-wave accelerator. This can then be used to electrically accelerate a hollow shell of strong material several microns in radius and containing about a cubic micron of cargo; such a shell can be given a high positive charge-to-mass ratio. Calculations indicate that an accelerator 1,000 kilometers long (there's room enough, in space) will be more than adequate to accelerate the shell and cargo to over 90 percent of the speed of light. A mass of one gram per meter for the accelerator (yielding a one-ton system) seems more than adequate. As the accelerator plunges through the target star system, it fires *backward* at a speed chosen to leave the cargo almost at rest. (For a discussion of the electrostatic acceleration of small particles, see "Impact Fusion and the Field Emission Projectile," by E. R. Harrison *[Nature,* Vol. 291, pp. 472–73, June 11, 1981].)

The residual velocity of the projectile can be directed to make it strike the atmosphere of a Mars- or Venus-like planet (selected beforehand by means of a large space-based telescope). A thin shell of the sort described will radiate atmospheric entry heat rapidly enough to remain cool. The cargo, consisting of an assembler and nanocomputer system, can then use the light of the local sun and local carbon, hydrogen, nitrogen, and oxygen (likely to be found in any planetary atmosphere) to replicate and to build larger structures.

An early project would be construction of a receiver for further instructions from home, including plans for complex devices. These can include rockets able to get off the planet (used as a target chiefly for its atmospheric cushion) to reach a better location for construction work. The resulting system of replicating assemblers could build virtually anything, including intelligent systems for exploration. To solve the lightsail stopping problem for the massive passenger vehicles that might follow, the system could build an array of braking lasers as large as the launching lasers back home. Their construction could be completed in a matter of weeks following delivery of the cubic-micron "seed." This system illustrates one way to spread human civilization to the stars at only slightly less than the speed of light.

P. 96 . . . **space near Earth holds** . . . Two days' travel at one gee acceleration can carry a person from Earth to any point on a disk having over 20 million times the area of Earth—and this calculation allows for a hole in the middle of the disk with a radius a hundred times the Earth-Moon distance. Even so, the outer edge of the disk reaches only one twentieth of the way to the Sun.

P. 96 . . . **enough energy in ten minutes** . . . Assuming conversion of solar to kinetic energy with roughly 10 percent efficiency, which should be achievable in any of several ways.

CHAPTER 7: ENGINES OF HEALING

P. 100 . . . **Dr. Seymour Cohen . . . argues . . .** See his article "Comparative Biochemistry and Drug Design for Infectious Disease" *(Science,* Vol. 205, pp. 964–71, Sept. 7, 1979).

P. 101 . . . **Researchers at Upjohn Company . . .** See "A Conformationally Constrained Vasopressin Analog with Antidiuretic Antagonistic Activity," by Gerald Skala et al. *(Science,* Vol. 226, pp. 443–45, Oct. 26, 1984).

P. 102 . . . **a dictionary definition of holism . . .** *The American Heritage Dictionary of the English Language,* edited by William Morris (Boston: Houghton Mifflin Company, 1978).

P. 104 . . . **aided by sophisticated technical AI systems . . .** These will be used both to help design molecular instruments and to direct their use. Using devices able to go to specified locations, grab molecules, and analyze them, the study of cell structures will become fairly easy to automate.

P. 105 . . . **separated molecules can be put back together . . .** Repair machines could use devices resembling the robots now used in industrial assembly work. But reassembling cellular structures will not require machines so precise (that is, so precise *for their size).* Many structures in cells will self-assemble if their components are merely confined together with freedom to bump around; they need not be manipulated in a complex and precise fashion. Cells already contain all the tools needed to assemble cell structures, and none is as complex as an industrial robot.

P. 105 . . . **the T4 phage . . . self-assembles . . .** See pp. 1022–23 of *Biochemistry,* by Albert L. Lehninger (New York: Worth Publishers, 1975).

P. 107 . . . **lipofuscin . . . fills over ten percent . . .** Lipofuscin contents vary with cell type, but some brain cells (in old animals) contain an average of about 17 percent; typical lipofuscin granules are one to three microns across. See "Lipofuscin Pigment Accumulation as a Function of Age and Distribution in Rodent Brain," by William Reichel et al. *(Journal of Gerontology,* Vol. 23, pp. 71–81, 1968). See also "Lipoprotein Pigments —Their Relationship to Aging in the Human Nervous System," by D. M. A. Mann and P. O. Yates *(Brain,* Vol. 97, pp. 481–88, 1974).

P. 107 . . . **about one in a million million million . . .** The implied relationship is not exact but shows the right trend: for example, the second number should be *2.33* uncorrected errors in a million million, and the third should be *4.44* in a million million million (according to some fairly complex calculations based on a slightly more complex correction algorithm).

P. 107 . . . **compare DNA molecules . . . make corrected copies . . .** Immune cells that produce different antibodies have different genes, edited during development. Repairing these genes will require special rules (but the demonstrated feasibility of growing an immune system shows that the right patterns of information can be generated).

P. 108 . . . **will identify molecules in a similar way** . . . Note that any molecule damaged enough to have an abnormal effect on the molecular machinery of the cell will by the same token be damaged enough to have a distinctive effect on molecular sensors.

P. 108 . . . **a complex and capable repair system** . . . For a monograph that discusses this topic in more detail, including calculations of volumes, speeds, powers, and computational loads, see "Cell Repair Systems," by K. Eric Drexler (available through The Foresight Institute, Palo Alto, Calif.).

P. 108 . . . **will be in communication** . . . For example, by means of hollow fibers a nanometer or two in diameter, each carrying a carbyne signaling rod of the sort used inside mechanical nanocomputers. Signal repeaters can be used where needed.

P. 114 . . . **to map damaged cellular structures** . . . This need not require solving any very difficult pattern recognition problems, save in cases where the cell structure is grossly disrupted. Each cell structure contains standard types of molecules in a pattern that varies within stereotyped limits, and a simple algorithm can identify even substantially-damaged proteins. Identification of the standard molecules in a structure determines its type; mapping it then becomes a matter of filling in known sorts of details.

P. 114 . . . **in a single calendar year** . . . Molecular experiments can be done about a millionfold faster than macroscopic experiments, since an assembler arm can perform actions at a million times the rate of a human arm (see Chapter 4). Thus, molecular machines and fast AI systems are well matched in speed.

P. 115 . . . **extended** . . . **lifespan** . . . **by 25 to 45 percent** . . . using 2-MEA, BHT, and ethoxyquine; results depended on the strain of mouse, the diet, and the chemical employed. See "Free Radical Theory of Aging," by D. Harman (*Triangle*, Vol. 12, No. 4, pp. 153–58, 1973).

P. 115 . . . **Eastman Kodak** . . . according to the *Press-Telegram*, Long Beach, Calif., April 26, 1985.

P. 115 . . . **rely on new science** . . . Cell repair will also rely on new science, but in a different way. As discussed in Chapter 3, it makes sense to predict what we will learn *about*, but not *what* we will learn. To extend life by means of cell repair machines will require that we learn *about* cell structures before we repair them, but *what* we learn will not affect the feasibility of those repairs. To extend life by conventional means, in contrast, will depend on how well the molecular machinery of the body can repair itself when properly treated. We will learn more about this, but what we learn could prove discouraging.

CHAPTER 8: LONG LIFE IN AN OPEN WORLD

P. 119 . . . durability has costs . . . See "Evolution of Aging," a review article by T. B. L. Kirkwood *(Nature,* Vol. 270, pp. 301–4, 1977).

P. 120 . . . As Sir Peter Medawar points out . . . in *The Uniqueness of the Individual* (London: Methuen, 1957). See also the discussion in *The Selfish Gene,* pp. 42–45 (in Chapter 2 references).

P. 120 . . . Experiments by Dr. Leonard Hayflick . . . See the reference above, which includes an alternative (but broadly similar) explanation for Hayflick's result.

P. 120 . . . A mechanism of this sort . . . For a reference to a statement of this theory (by D. Dykhuizen in 1974) together with a criticism and a rebuttal, see the letters by Robin Holliday, and by John Cairns and Jonathan Logan, under "Cancer and Cell Senescence" *(Nature,* Vol. 306, p. 742, December 29, 1983).

P. 120 . . . could harm older animals by stopping . . . division . . . These animals could still have a high cancer rate because of a high incidence of broken clocks.

P. 121 . . . cleaning machines to remove these poisons . . . One system's meat really is another's poison; cars "eat" a toxic petroleum product. Even among organisms, some bacteria thrive on a combination of methanol (wood alcohol) and carbon monoxide (see "Single-Carbon Chemistry of Acetogenic and Methanogenic Bacteria," by J. G. Zeikus et al., *Science,* Vol. 227, pp. 1167–71, March 8, 1985), while others have been bred that can live on either trichlorophenol or the herbicide 2,4,5-T. They can even defluorinate pentafluorophenol. (See "Microbial Degradation of Halogenated Compounds," D. Ghousal et al., *Science,* Vol. 228, pp. 135–42, April 12, 1985.)

P. 121 . . . cheap enough to eliminate the need for fossil fuels . . . through the use of fuels made by means of solar energy.

P. 121 . . . able to extract carbon dioxide from the air . . . In terms of sheer tonnage, carbon dioxide is perhaps our biggest pollution problem. Yet, surprisingly, a simple calculation shows that the sunlight striking Earth in a day contains enough energy to split all the carbon dioxide in the atmosphere into carbon and oxygen (efficiency considerations aside). Even allowing for various practical and aesthetic limitations, we will have ample energy to complete this greatest of cleanups in the span of a single decade.

P. 122 . . . Alan Wilson . . . and his co-workers . . . See "Gene Samples from an Extinct Animal Cloned," by J. A. Miller *(Science News,* Vol. 125, p. 356, June 9, 1984).

P. 125 . . . O my friend . . . *The Iliad,* by Homer (about the eighth century B.C.), as quoted by Eric Hoffer in *The True Believer* (New York: Harper & Brothers, 1951). (Sarpedon is indeed killed in the battle.)

P. 126 . . . **Gilgamesh, King of Uruk** . . . From *The Epic of Gilgamesh*, translated by N. K. Sandars (Middlesex: Penguin Books, 1972).

CHAPTER 9: A DOOR TO THE FUTURE

P. 130 . . . **To Jacques Dubourg** . . . In *Mr. Franklin, A Selection from His Personal Letters*, by L. W. Labaree and W. J. Bell, Jr. (New Haven: Yale University Press, 1956), pp. 27–29.

P. 131 . . . **a new heart, fresh kidneys, or younger skin** . . . With organs and tissues grown from the recipient's own cells, there will be no problem of rejection.

P. 132 . . . **The changes . . . are far from subtle** . . . Experiments show that variations in experience rapidly produce visible variations in the shape of dendritic spines (small synapse-bearing protrusions on dendrites). See "A Theoretical Analysis of Electrical Properties of Spines," by C. Koch and T. Poggio (MIT AI Lab Memo No. 713, April 1983).

In "Cell Biology of Synaptic Plasticity" *(Science,* Vol. 225, pp. 1287–94, Sept. 21, 1984), Carl W. Cotman and Manuel Nieto-Sampedro write that "The nervous system is specialized to mediate the adaptive response of the organism. . . . To this end the nervous system is uniquely modifiable, or plastic. Neuronal plasticity is largely the capability of synapses to modify their function, to be replaced, and to increase or decrease in number when required." Further, "because the neocortex is believed to be one of the sites of learning and memory, most of the studies of the synaptic effect of natural stimuli have concentrated on this area." Increases in dendritic branching in the neocortex "are caused by age (experience) in both rodents and humans. Smaller but reproducible increases are observed after learning of particular tasks . . ." These changes in cell structure can occur "within hours."

For a discussion of short- and long-term memory, and of how the first may be converted into the second, see "The Biochemistry of Memory: A New and Specific Hypothesis," by Gary Lynch and Michel Baudry *(Science,* Vol. 224, pp. 1057–63, June 8, 1984).

At present, every viable theory of long-term memory involves changes in the structure and protein content of neurons. There is a persistent popular idea that memory might somehow be stored (exclusively?) "in RNA molecules," a rumor seemingly fostered by an analogy with DNA, the "memory" responsible for heredity. This idea stems from old experiments suggesting that learned behaviors could be transferred to uneducated flatworms by injecting them with RNA extracted from educated worms. Unfortunately for this theory, the same results were obtained using RNA from entirely uneducated yeast cells. See *Biology Today,* by David Kirk, p. 616 (New York: Random House, 1975).

Another persistent popular idea is that memory might be stored in the form of reverberating patterns of electrical activity, a rumor seemingly fostered by an analogy with the dynamic random-access memories of modern computers. This analogy, however, is inappropriate for several reasons: (1) Computer memories, unlike brains, are designed to be erased and reused repeatedly. (2) The patterns in a computer's "long-term memory"—its magnetic disk, for example—are in fact more durable than dynamic RAM. (3) Silicon chips are designed for structural stability, while the brain is designed for dynamic structural change. In light of the modern evidence for long-term memory storage in long-lasting brain structures, it is not surprising that "total cessation of the electrical activity of the brain does not generally delete memories, although it may selectively affect the most recently stored ones" (A. J. Dunn). The electrical reverberation theory was proposed by R. Lorente de Nó *(Journal of Neurophysiology,* Vol. 1, p. 207) in *1938.* Modern evidence fails to support such theories of ephemeral memory.

P. 132 . . . "striking morphological changes" . . . For technical reasons, this study was performed in mollusks, but neurobiology has proved surprisingly uniform. See "Molecular Biology of Learning: Modulation of Transmitter Release," by Eric R. Kandel and James H. Schwartz *(Science,* Vol. 218, pp. 433–43, Oct. 29, 1982), which reports work by C. Baily and M. Chen.

P. 133 . . . until after vital functions have ceased . . . The time between expiration and dissolution defines the window for successful biostasis, but this time is uncertain. As medical experience shows, it is possible to destroy the brain (causing irreversible dissolution of mind and memory) even while a patient breathes. In contrast, patients have been successfully revived after a significant period of so-called "clinical death." With cell repair machines, the basic requirement is that brain cells remain structurally intact; so long as they are *alive,* they are presumably intact, so viability provides a conservative indicator.

There is a common myth that the brain "cannot survive" for more than a few minutes without oxygen. Even if this were true regarding survival of the (spontaneous) ability to resume *function,* the survival of characteristic cell *structure* would still be another matter. And indeed, cell structures in the brains of expired dogs, even when kept at room temperature, show only moderate changes after six hours, and many cell structures remain visible for a day or more; see "Studies on the Epithalamus," by Duane E. Haines and Thomas W. Jenkins *(Journal of Comparative Neurology,* Vol. 132, pp. 405–17, Mar. 1968).

But in fact, the potential for spontaneous brain function can survive for longer than this myth (and the medical definition of "brain death") would suggest. A variety of experiments employing drugs and surgery show this: Adult monkeys have completely recovered after a sixteen-minute cutoff of circulation to the brain (a condition, called "ischemia," which clearly blocks oxygen supply as well); see "Thiopental Therapy After 16 Minutes of Global Brain Ischemia in Monkeys, by A. L. Bleyaert et al. *(Critical Care Medicine,* Vol. 4, pp. 130–31, Mar./Apr. 1976). Monkey and cat brains have survived for an hour at body temperature without circulation, then recovered electrical function; see "Reversibility of Ischemic Brain Damage," by K.-A. Hossmann and Paul Kleihues *(Archives of Neurology,* Vol. 29, pp. 375–84, Dec. 1973). Dr. Hossmann concludes that "any nerve cell in the brain can survive" for an hour without blood (after the heart stops pumping, for example). The problem is not that nerve cells die when circulation stops, but that secondary problems (such as a slight swelling of the brain within its tight-fitting bone case) can prevent circulation from resuming. When chilled to near freezing, dog brains have recovered electrical activity after four hours without circulation (and have recovered substantial metabolic activity even after *fifteen days);* see "Prolonged Whole-Brain Refrigeration with Electrical and Metabolic Recovery," by Robert J. White et al. *(Nature,* Vol. 209, pp. 1320–22, Mar. 26, 1966).

Brain cells that retain the capability for *spontaneous* revival at the time when they undergo biostasis should prove easy to repair. Since success chiefly requires that characteristic cell structures remain intact, the time window for beginning biostasis procedures is probably at least several hours after expiration, and possibly longer. Cooperative hospitals can and have made the time much shorter.

P. 133 . . . **fixation procedures preserve cells** . . . For high-voltage electron micrographs showing molecular-scale detail in cells preserved by glutaraldehyde fixation, see "The Ground Substance of the Living Cell," by Keith R. Porter and Jonathan B. Tucker *(Scientific American,* Vol. 244, pp. 56–68, Mar. 1981). Fixation alone does not seem sufficient; long-term stabilization of structure seems to demand freezing or vitrification, either alone or in addition to fixation. Cooling in nitrogen—to minus 196 degrees C—can preserve tissue structures for many thousands of years.

P. 134 . . . **solidification without freezing** . . . See "Vitrification as an Approach to Cryopreservation," by G. M. Fahy et al. *(Cryobiology,* Vol. 21, pp. 407–26, 1984).

P. 134 . . . **Mouse embryos** . . . See "Ice-free Cryopreservation of Mouse Embryos at −196° C by Vitrification," by W. F. Rall and G. M. Fahy *(Nature,* Vol. 313, pp. 573–75, Feb. 14, 1985).

P. 134 . . . **Robert Ettinger** . . . published a book . . . *The Prospect of Immortality* (New York: Doubleday, 1964; a preliminary version was privately published in 1962).

P. 135 . . . many human cells revive *spontaneously* . . . It is well known that human sperm cells and early embryos survive freezing and storage; in both cases, successes have been reported in the mass media. Less spectacular successes with other cell types (frozen and thawed blood is used for transfusions) are numerous. It is also interesting to note that, after treatment with glycerol and freezing to minus 20 degrees C, cat brains can recover spontaneous electrical activity after over 200 days of storage; see "Viability of Long Term Frozen Cat Brain *In Vitro,*" by I. Suda, K. Kito, and C. Adachi *(Nature,* Vol. 212, pp. 268–70, Oct. 15, 1966).

P. 135 . . . researching ways to freeze and thaw viable organs . . . A group at the Cryobiology Laboratory of The American Red Cross (9312 Old Georgetown Road, Bethesda, Md. 20814) is pursuing the preservation of whole human organs to allow the establishment of banks of organs for transplantation; see "Vitrification as an Approach to Cryopreservation," referenced above.

P. 135 . . . cell repair . . . has been a consistent theme . . . As I found when the evident feasibility of cell repair finally led me to examine the cryonics literature. Robert Ettinger's original book, for example (referenced above), speaks of the eventual development of "huge surgeon machines" able to repair tissues "cell by cell, or even molecule by molecule in critical areas." In 1969 Jerome B. White gave a paper on "Viral Induced Repair of Damaged Neurons with Preservation of Long Term Information Content," proposing that means might be found to direct repair using artificial viruses; see the abstract quoted in *Man into Superman,* by Robert C. W. Ettinger (New York: St. Martin's Press, 1972, p. 298). In "The Anabolocyte: A Biological Approach to Repairing Cryoinjury" *(Life Extension Magazine,* pp. 80–83, July/August 1977), Michael Darwin proposed that it might be possible to use genetic engineering to make highly modified white blood cells able to take apart and reconstruct damaged cells. In "How Will They Bring Us Back, 200 Years From Now?" *(The Immortalist,* Vol. 12, pp. 5–10, Mar. 1981), Thomas Donaldson proposed that systems of molecular machines (with devices as small as viruses and aggregates of devices as large as buildings, if need be) could perform any needed repairs on frozen tissues.

The idea of cell repair systems has thus been around for many years. The concepts of the assembler and the nanocomputer have now made it possible to see clearly how such devices can be built and controlled, and that they can in fact fit within cells.

P. 135 . . . the animals fail to revive . . . Hamsters, however, have been cooled to a temperature which froze over half the water content in their bodies (and brains), and have then revived with complete recovery; see *Biological Effects of Freezing and Supercooling,* by Audrey U. Smith (Baltimore: Williams & Wilkins, 1961).

P. 135 . . . **As Robert Prehoda stated** . . . in *Designing the Future: The Role of Technological Forecasting* (Philadelphia: Chilton Book Co., 1967).

P. 136 . . . **discouraged the use of a workable biostasis technique** . . . Other factors have also been discouraging—chiefly cost and ignorance. For a patient to pay for a biostasis procedure and to establish a fund that provides for indefinite storage in liquid nitrogen now costs $35,000 or more, depending on the biostasis procedure chosen. This cost is typically covered by purchasing a suitable life insurance policy. Facing this cost and having no clear picture of how freezing damage can be repaired, only a few patients out of millions have so far chosen this course. The small demand, in turn, has prevented economies of scale from lowering the cost of the service. But this may be about to change. Cryonics groups report a recent increase in biostasis contracts, apparently stemming from knowledge of advances in molecular biology and in the understanding of future cell repair capabilities.

Three U.S. groups presently offer biostasis services. In order of their apparent size and quality, they are:

The Alcor Life Extension Foundation, 4030 North Palm No. 304, Fullerton, Calif. 92635, (714) 738-5569. (Alcor also has a branch and facilities in southern Florida.)

Trans Time, Inc., 1507 63rd Street, Emeryville, Calif. 94707, (415) 655-9734.

The Cryonics Institute, 24041 Stratford, Oak Park, Mich. 48237, (313) 967-3115.

For practical reasons based on experience, they require that legal and financial arrangements be completed in advance.

P. 136 . . . **this preserves neural structures** . . . The growth of ice crystals can displace cell structures by a few millionths of a meter, but it does not obliterate them, nor does it seem likely to cause any significant confusion regarding where they were before being displaced. Once frozen, they move no further. Repairs can commence before thawing lets them move again.

P. 137 . . . **clearing** . . . **the major blood vessels** . . . Current biostasis procedures involve washing out most of a patient's blood; the nanomachines recover any remaining blood cells as they clear the circulatory system.

P. 137 . . . **throughout the normally active tissues** . . . This excludes, for example, the cornea, but other means can be used to gain access to the interior of such tissues, or they can simply be replaced.

P. 137 . . . **that enter cells and remove the glassy protectant** . . . Molecules of protectant are bound to one another by bonds so weak that they break at room temperature from thermal vibrations. Even at low temperatures, protectant-removal machines will have no trouble pulling these molecules loose from surfaces.

P. 137 . . . **a temporary molecular scaffolding** . . . This could be built of nanometer-thick rods, designed to snap together. Molecules could be fastened to the scaffolding with devices resembling double-ended alligator clips.

P. 137 . . . **the machines label them** . . . Labels can be made from small segments of coded polymer tape. A segment a few nanometers long can specify a location anywhere within a cubic micron to one-nanometer precision.

P. 137 . . . **report** . . . **to a larger computer within the cell** . . . In fact, a bundle of nanometer-diameter signal-transmission fibers the diameter of a finger (with slender branches throughout the patient's capillaries) can in less than a week transmit a complete molecular description of all a patient's cells to a set of external computers. Though apparently unnecessary, the use of external computers would remove most of the significant volume, speed, and power-dissipation constraints on the amount of computation available to plan repair procedures.

P. 137 . . . **identifies cell structures from molecular patterns** . . . Cells have stereotyped structures, each built from standard kinds of molecules connected in standard ways in accordance with standard genetic programs. This will greatly simplify the identification problem.

P. 141 . . . **Richard Feynman saw** . . . He pointed out the possibility of making devices with wires as little as ten or a hundred atoms wide; see "There's Plenty of Room at the Bottom," in *Miniaturization,* edited by H. D. Gilbert (New York: Reinhold, 1961), pp. 282–96.

P. 142 . . . **Robert T. Jones wrote** . . . in "The Idea of Progress" *(Astronautics and Aeronautics,* p. 60, May 1981).

P. 142 . . . **Dr. Lewis Thomas wrote** . . . in "Basic Medical Research: A Long-Term Investment" *(Technology Review,* pp. 46–47, May/June 1981).

P. 143 . . . **Joseph Lister published** . . . See Volume V, "Fine Chemicals" in *A History of Technology,* edited by C. J. Singer and others (Oxford: Clarendon Press, 1958).

P. 144 . . . **Sir Humphry Davy wrote** . . . See *A History of Technology,* referenced above.

CHAPTER 10: THE LIMITS TO GROWTH

P. 149 . . . **the limiting speed is nothing so crude or so breakable** . . . The principle of relativity of motion means that "moving" objects may be considered to be at rest—meaning that a spaceship pilot trying to approach the speed of light wouldn't even know in what direction to accelerate. Further, simple Minkowski diagrams show that the geometry of space-time makes traveling faster than light equivalent to traveling backward in time —and where do you point a rocket to move in *that* direction?

P. 149 . . . **Arthur C. Clarke wrote** . . . in *Profiles of the Future: An Inquiry into the Limits of the Possible,* first edition (New York: Harper & Row, 1962).

P. 151 . . . **its properties limit all that we can do** . . . For an account of some modern theories that attempt to unify all physics in terms of the behavior of the vacuum, see "The Hidden Dimensions of Spacetime," by Daniel Z. Freedman and Peter van Nieuwenhuizen *(Scientific American,* Vol. 252, pp. 74–81, Mar. 1985).

P. 152 . . . **peculiarities far more subtle** . . . For example, quantum measurements can affect the outcome of other quantum measurements instantaneously at an arbitrarily great distance—but the effects are only statistical and of a subtle sort that has been mathematically proved to be unable to transmit information. See the very readable discussion of Bell's theorem and the Einstein-Podolsky-Rosen paradox in *Quantum Reality* by Nick Herbert (Garden City, New York: Anchor Press/Doubleday, 1985). Despite rumors to the contrary (some passed on in the final pages of *Quantum Reality),* nothing seems to suggest that consciousness and the mind rely on quantum mechanics in any special way. For an excellent discussion of how consciousness works (and of how little consciousness we really have) see Marvin Minsky's *The Society of Mind* (New York: Simon & Schuster, 1986).

P. 153 . . . **Your victim might have said something vague** . . . But now physics can answer those questions with clear mathematics. Calculations based on the equations of quantum mechanics show that air is gaseous because nitrogen and oxygen atoms tend to form tightly bonded pairs, unbonded to anything else. Air is transparent because visible light lacks enough energy to excite its tightly bound electrons to higher quantum states, so photons pass through without absorption. A wooden desk is solid because it contains carbon atoms which (as shown by quantum mechanical calculations) are able to form the tightly bonded chains of cellulose and lignin. It is brown because its electrons are in a variety of states, some able to be excited by visible light; it preferentially absorbs bluer, higher-energy photons, making the reflected light yellowish or reddish.

P. 153 . . . **Stephen W. Hawking states** . . . In "The Edge of Space-time" *(American Scientist,* Vol. 72, pp. 355–59, Jul.–Aug. 1984).

P. 154 . . . **Few other stable particles are known** . . . Electrons, protons, and neutrons have stable antiparticles with virtually identical properties save for opposite charges and the ability to annihilate when paired with their twins, releasing energy (or lighter particles). They thus have obvious applications in energy storage. Further, antimatter objects (made from the antiparticles of ordinary matter) may have utility as negative electrodes in high-field electrostatic systems: the field would have no tendency to remove positrons (as it would electrons), making mechanical disruption of the electrode surface the chief limit to field strength. Such electrodes would have to be made and positioned without contacting ordinary matter, of course.

Various physical theories predict a variety of other stable particles (and even massive, particle-like lines), but all would be either so weakly interacting as to be almost undetectable (like neutrinos, only more so) or very massive (like hypothesized magnetic monopoles). Such particles could still be very useful, if found.

P. 155 . . . Trying to change a nucleus . . . The molecular and field effects used in nuclear magnetic resonance spectroscopy change the orientation of a nucleus, but not its structure.

P. 155 . . . the properties of well-separated nuclei . . . It has been suggested that excited nuclei might even be made to serve as the lasing medium in a gamma-ray laser.

P. 156 . . . would present substantial difficulties . . . Before nuclei are pushed close enough together to interact, the associated atomic structures merge to form a solid, metal-like "degenerate matter," stable only under enormous pressure. When the nuclei finally do interact, the exchange of neutrons and other particles soon transmutes them all into similar kinds, obliterating many of the patterns one might seek to build and use.

P. 156 . . . insulating against heat . . . This is a simple goal to state, but the optimal structures (at least where some compressive strength is required) may be quite complex. Regular crystals transmit heat well, making irregularity desirable, and irregularity means complexity.

P. 157 . . . the runners-up will often be nearly as good . . . And in some instances, we may design the best possible system, yet never be sure that better systems do not exist.

P. 157 . . . Richard Barnet writes . . . in *The Lean Years: Politics in the Age of Scarcity* (New York: Simon & Schuster, 1980).

P. 158 . . . Jeremy Rifkin (with Ted Howard) has written . . . *Entropy: A New World View* (New York: Viking Press, 1980).

P. 160 . . . "The ultimate moral imperative, then . . ." Despite this statement, Rifkin has since struck off on a fresh moral crusade, this time against the idea of evolution and against *human beings'* modifying genes, even in ways that viruses and bacteria have done for millions of years. Again, he warns of cosmic consequences. But he apparently still believes in the tightly sealed, ever dying world he described in *Entropy:* "We live by the grace of sacrifice. Every amplification of our being owes its existence to some diminution somewhere else." Having proved in *Entropy* that he misunderstands how the cosmos works, he now seeks to advise us about what it *wants:* "The interests of the cosmos are no different from ours. . . . How then do we best represent the interests of the cosmos? By paying back to the extent to which we have received." But he seems to see all human achievements as fundamentally destructive, stating that "the only living legacy that we can ever leave is the endowment we never touched," and declaring that "life requires death." For more misanthropy and misconceptions, see *Algeny,* by Jeremy Rifkin (New York: Viking, 1983).

For a confident assertion that genetic engineering is impossible in the first place, made by Rifkin's "prophet and teacher," Nicholas Georgescu-Roegen, see *The Entropy Law and the Economic Process* (Cambridge, Mass.: Harvard University Press, 1971).

P. 163 . . . **exponential growth will overrun** . . . The demographic transition—the lowering of average birthrates with economic growth—is basically irrelevant to this. The exponential growth of even a tiny minority would swiftly make it a majority, and then make it consume all available resources.

P. 164 . . . **exploding outward at near the speed of light** . . . The reason for this rests on a very basic evolutionary argument. Assume that a diverse, competitive civilization begins expanding into space. What groups will be found at the frontier? Precisely those groups that expand fastest. The competition for access to the frontier provides an evolutionary pressure that favors maximum speed of travel and settlement, and that maximum speed is little short of the speed of light (see the notes to Chapter 6). In a hundred million years, such civilizations would spread not just across galaxies, but across intergalactic space. That a thousand or a million times as many civilizations might collapse before reaching space, or might survive without expanding, is simply irrelevant. A fundamental lesson of evolution is that, where replicators are concerned, a single success can outweigh an unlimited number of failures.

P. 165 . . . **need not contain every possible chemical** . . . Even the number of possible DNA molecules 50 nucleotides long (four to the fiftieth power) is greater than the number of molecules in a glass of water.

P. 166 . . . *The Limits to Growth* . . . by Donella H. Meadows et al. (New York: Universe Books, 1972).

P. 166 . . . *Mankind at the Turning Point* . . . by Mihajlo D. Mesarović and Eduard Pestel (New York: Dutton, 1974).

CHAPTER 11: ENGINES OF DESTRUCTION

P. 172 . . . **trouble enough controlling viruses and fruit flies** . . . We have trouble even though they are made of conventional molecular machinery. Bacteria are also hard to control, yet they are superficially almost helpless. Each bacterial cell resembles a small, rigid, mouthless box—to eat, a bacterium must be immersed in a film of water that can carry dissolved nutrients for it to absorb. In contrast, assembler-based "superbacteria" could work with or without water; they could feed their molecular machinery with raw materials collected by "mouths" able to attack solid structures.

P. 174 . . . AI systems could serve as . . . strategists, or fighters . . . See "The Fifth Generation: Taking Stock," by M. Mitchell Waldrop *(Science,* Vol. 226, pp. 1061–63, Nov. 30, 1984), and "Military Robots," by Joseph K. Corrado *(Design News,* pp. 45–66, Oct. 10, 1983).

P. 178 . . . none, if need be . . . To be precise, an object can be assembled with a negligible chance of putting any atoms in the wrong place. During assembly, errors can be made arbitrarily unlikely by a process of repeated testing and correction (see the notes for Chapter 4). For example, assume that errors are fairly common. Assume further that a test sometimes fails, allowing one in every thousand errors to pass undetected. If so, then a series of twenty tests will make the chance of failing to detect and correct an error so low that the odds of misplacing a single atom would be slight, even in making an object the size of the Earth. But radiation damage (occurring at a rate proportional to the object's size and age) will eventually displace even correctly placed atoms, so this degree of care would be pointless.

P. 178 . . . a cosmic ray can unexpectedly knock atoms loose from anything . . . It might seem that shielding could eliminate this problem, but neutrinos able to penetrate the entire thickness of the Earth—or Jupiter, or the Sun—can still cause radiation damage, though at a very small rate. See "The Search for Proton Decay," by J. M. LoSecco et al. *(Scientific American,* Vol. 252, p. 59, June 1985).

P. 178 . . . Stratus Computer Inc., for example . . . See "Fault-Tolerant Systems in Commercial Applications," by Omri Serlin *(Computer,* Vol. 17, pp. 19–30, Aug. 1984).

P. 179 . . . design diversity . . . See "Fault Tolerance by Design Diversity: Concepts and Experiments," by Algirdas Avižienis and John P. J. Kelly *(Computer,* Vol. 17, pp. 67–80, Aug. 1984).

P. 179 . . . redundancy . . . multiple DNA strands . . . The bacterium *Micrococcus radiodurans* apparently has quadruple-redundant DNA, enabling it to survive extreme radiation doses. See "Multiplicity of Genome Equivalents in the Radiation-Resistant Bacterium *Micrococcus radiodurans,*" by Mogens T. Hansen, in *Journal of Bacteriology,* pp. 71–75, Apr. 1978.

P. 179 . . . other effective error-correcting systems . . . Error correction based on multiple copies is easier to explain, but digital audio disks (for example) use other methods that allow error correction with far less redundant information. For an explanation of a common error-correcting code, see "The Reliability of Computer Memory," by Robert McEliece *(Scientific American,* Vol. 248, pp. 88–92, Jan. 1985).

P. 180 . . . intelligence will involve mental parts . . . See *The Society of Mind,* by Marvin Minsky (New York: Simon & Schuster, 1986).

P. 181 . . . "The Scientific Community Metaphor" . . . by William A. Kornfeld and Carl Hewitt (MIT AI Lab Memo No. 641, Jan. 1981).

P. 181 . . . AI systems can be made trustworthy . . . Safety does not require that *all* AI systems be made trustworthy, so long as *some* are trustworthy and help us plan precautions for the rest.

P. 181 . . . One proposal . . . This is a concept being developed by Mark Miller and myself; it is related to the ideas discussed in "Open Systems," by Carl Hewitt and Peter de Jong (MIT AI Lab Memo No. 692, Dec. 1982).

P. 181 . . . more reliably than . . . human engineers . . . if only because fast AI systems (like those described in Chapter 5) will be able to find and correct errors a million times faster.

P. 183 . . . other than specially designed AI programs . . . See "The Role of Heuristics in Learning by Discovery," by Douglas B. Lenat, in *Machine Learning,* edited by Michalski et al. (Palo Alto, Calif.: Tioga Publishing Company, 1983). For a discussion of the successful evolution of programs designed to evolve, see pp. 243–85. For a discussion of the unsuccessful evolution of programs intended to evolve but not properly designed to do so, see pp. 288–92.

P. 183 . . . neglect to give replicators similar talents . . . Lacking these, "gray goo" might be able to replace us and yet be unable to evolve into anything interesting.

P. 185 . . . to correct its calculations . . . Calculations will allow the system to picture molecular structures that it has not directly characterized. But calculations may lead to ambiguous results in borderline cases—actual results may even depend on random tunneling or thermal noise. In this case, the measurement of a few selected atomic positions (performed by direct mechanical probing of the workpiece's surface) should suffice to distinguish among the possibilities, thus correcting the calculations. This can also correct for error buildup in calculations of the geometry of large structures.

P. 185 . . . Each sensor layer . . . As described, these sensor layers must be penetrated by wires, which might seem to present a security problem: what if something were to get past the sensors by eating its way along a wire? In practice, anything that can transmit signals and power (including optical fibers, mechanical transmission systems, and so forth) could be used in place of wires. These channels can be made secure by basing them on materials with extreme properties: if a very fine wire is made of the most conductive material, or if a mechanical transmission system is made of the strongest material (and used near its breaking stress), then any attempt to replace a segment with something else (such as an escaping replicator) will show up as a greater electrical resistance or a fractured part. Thus the transmission systems themselves can act as sensors. For the sake of redundancy and design diversity, different sensor layers could be penetrated by different transmission systems, each transmitting signals and power to the next.

P. 186 . . . If we destroy the records of the protein designs . . . But how could people be made to forget? This is not really necessary, since their knowledge would be dispersed. In developing modern hardware systems, different teams work on different parts; they need not know what another team's part *is* (much less how it is made), because only how it *interacts* really matters to them. In this way people reached the Moon, though no one person or team ever fully knew how to do it; it could be likewise with assemblers.

Since the first assembler designs will be historic documents, it might be better to store them securely, rather than destroy them. Eventually they can become part of the open literature. But hiding design information will at best be a stopgap, since the *methods* used for the design of the first assembler system will be harder to keep secret. Further, sealed assembler labs might be used to develop and test machines that can make assemblers, even machines that can themselves be made *without* assemblers.

P. 187 . . . no fixed wall will be proof against large-scale, organized malice . . . Sealed assembler labs can work despite this. They do not protect their contents from the outside; in fact, they are designed to destroy their contents when tampered with. Instead, they protect the outside from their contents—and their sealed work spaces are too small to hold any large-scale system, malicious or not.

P. 188 . . . giving the attacker no obvious advantage . . . In the examples cited, organized entities were pitted against similar entities. These entities could, of course, be vaporized by hydrogen bombs, but faced with the prospect of retaliation in kind, no attacker has yet seen an advantage in launching a nuclear strike.

CHAPTER 12: STRATEGIES AND SURVIVAL

P. 194 . . . occupy hostile powers . . . In principle, this could be a minimal form of occupation, controlling only research laboratories, but even this would require a degree of coercion roughly equivalent to conquest.

P. 197 . . . as open as possible . . . It may be possible to devise forms of inspection that give a group great confidence in what a system under development will (and will not) be able *to do,* without letting that group learn how those systems *are made.* Compartmentalized development of a system's components could, in principle, allow several groups to cooperate without any single group's being able to build and use a similar system independently.

P. 198 . . . we naturally picture human hands aiming it . . . For a discussion of autonomous spacecraft, see "Expanding Role for Autonomy in Military Space," by David D. Evans and Maj. Ralph R. Gajewski *(Aerospace America,* pp. 74–77, Feb. 1985). See also "Can Space Weapons Serve Peace?" by K. Eric Drexler *(L5 News,* Vol. 9, pp. 1–2, Jul. 1983).

P. 198 . . . while providing each with some protection . . . Saying that a symmetrical, 50 percent effective shield would be worthless is like saying that a bilateral 50 percent reduction in nuclear missiles—a real breakthrough in arms control—would be worthless. The practicality of such a shield is another matter. Until really good active shields become possible, the question is not one of making a nuclear attack harmless, but at best of making it less likely.

P. 199 . . . limiting technology transfer . . . In fact, President Reagan has spoken of giving away U.S. space defense technology to the Soviet Union. See the New York *Times,* p. A15, March 30, 1983. See also "Sharing 'Star Wars' technology with Soviets a distant possibility, says head of Pentagon study group," by John Horgan *(The Institute,* p. 10, Mar. 1984). Richard Ullman, professor of international affairs at Princeton University, has proposed a joint defense program with extensive sharing of technology; see "U.N.-doing Missiles" (New York *Times,* p. A23, Apr. 28, 1983).

In principle, a joint project could proceed with little technology transfer. There is a great difference between (1) knowing *what a device cannot do,* (2) knowing *what it can do,* (3) knowing *what it is,* and (4) knowing *how to make it.* These define four levels of knowledge, each (more or less) independent of the levels beyond it. For example, if I were to hand you a plastic box, a superficial examination might convince you that it cannot fly or shoot bullets, but not tell you what it can do. A demonstration might then convince you that it can serve as a cordless telephone. By inspecting it more closely, you could trace its circuits and gain an excellent idea of what it is and of what its operating limits are. But you still wouldn't necessarily know how to make one.

The essence of an active shield lies in what it *cannot do*—that is, that it cannot be used as a weapon. To conduct a joint active-shield project relying on high-technology components, one would need to share knowledge chiefly on levels (1) and (2). This requires at least limited sharing on level (3), but need not require any on level (4).

P. 199 . . . basic issues common to all active shields . . . Such as those of their control, purpose, and reliability, and the fundamental issue of political understanding and acceptance.

CHAPTER 13: FINDING THE FACTS

P. 204 . . . a U.S. National Science Foundation survey . . . as quoted by NSF Director John B. Slaughter *(Time,* p. 55, June 15, 1981).

P. 204 . . . *Advice and Dissent* . . . subtitled *"Scientists in the Political Arena,"* by Joel Primack and Frank von Hippel (New York: Basic Books, 1974).

P. 204 . . . Hazel Henderson argues . . . in *Creating Alternative Futures: The End of Economics* (New York: Berkley Publishing, 1978).

P. 204 . . . Harrison Brown likewise argues . . . in *The Human Future Revisited: The World Predicament and Possible Solutions* (New York: Norton, 1978).

P. 205 . . . Debates . . . over the safety of nuclear power . . . For a discussion of the failures at the Three Mile Island nuclear power plant, and a discussion of (1) the remarkable degree of agreement on the problems reached by an expert panel and (2) how the media mangled the story, and (3) how the federal government had failed to respond to reality, see "Saving American Democracy," by John G. Kemeny, president of Dartmouth College and chairman of the presidential commission on Three Mile Island *(Technology Review,* pp. 65–75, June/July 1980). He concludes that "the present system does not work."

P. 206 . . . Disputes over facts . . . Worse yet, two people can agree on the facts *and* on basic values (say, that wealth is good and pollution is bad) and yet disagree about building a factory—one person may be more concerned about wealth, and the other about pollution. In emotional debates, this can lead each side to accuse the other of perverted values, such as favoring poverty or caring nothing for the environment. Nanotechnology will ease such conflicts by changing the trade-offs. Because we can have much more wealth *and* much less pollution, old opponents may more often find themselves in agreement.

P. 207 . . . AI researchers . . . See "The Scientific Community Metaphor," by William A. Kornfeld and Carl Hewitt (MIT AI Lab Memo No. 641, Jan. 1981); see also the discussion of "due-process reasoning" in "The Challenge of Open Systems," by Carl Hewitt *(Byte,* Vol. 10, pp. 223–41, April 1985).

P. 209 . . . procedures somewhat like those of courts . . . These might use written communications, as in journals, rather than face-to-face meetings: judging the truth of a statement by the manner in which it is said is useful in courts, but plays a lesser role in science.

P. 211 . . . Kantrowitz . . . originated the concept . . . He did so in the mid-1960s. See his discussion in "Controlling Technology Democratically" *(American Scientist,* Vol. 63, pp. 505–9, Sept.–Oct. 1975).

P. 211 . . . used (or proposed) as a government institution . . . This is the original usage of the term "science court," and many criticisms of the due-process idea have stemmed from this aspect of the proposal. The fact forum approach is genuinely different; Dr. Kantrowitz is presently pursuing it under the name "Scientific Adversary Procedure."

P. 211 . . . backed by the findings of an expert committee . . . which in 1960 had drawn up a proposed space program for the Air Force. It emphasized that learning to assemble systems in Earth orbit (such as space stations and Moon ships) was at least as important as building bigger boosters. During the subsequent debate on how to reach the Moon, Kantrowitz argued that Earth-orbital assembly would be perhaps ten times less expensive than the giant-rocket, lunar-orbit-rendezvous approach that was finally chosen. But political factors intervened, and the matter never received a proper public hearing. See "Arthur Kantrowitz Proposes a Science Court," an interview by K. Eric Drexler *(L5 News,* Vol. 2, p. 16, May 1977).

For an account of another abuse of technical decision-making during Apollo, see *The Heavens and the Earth: A Political History of the Space Age,* by William McDougall, pp. 315–16 (New York: Basic Books, 1985).

P. 211 . . . a proposed procedure . . . This is described in "The Science Court Experiment: An Interim Report," by the Task Force of the Presidential Advisory Group on Anticipated Advances in Science and Technology *(Science,* Vol. 193, pp. 653–56, Aug. 20, 1976).

P. 213 . . . a colloquium on the science court . . . see the *Proceedings of the Colloquium on the Science Court,* Leesburg, Virginia, Sept. 20–21, 1976 (National Technical Information Center, document number PB261305). For a summary and discussion of the criticisms voiced at the colloquium, see "The Science Court Experiment: Criticisms and Responses," by Arthur Kantrowitz *(Bulletin of the Atomic Scientists,* Vol. 33, pp. 44–49, Apr. 1977).

P. 214 . . . could move toward due process . . . The formation of the Health Effects Institute of Cambridge, Massachusetts, created in 1980 to bring together adversaries in the field of air pollution, has been a step in this direction. See "Health Effects Institute Links Adversaries," by Eliot Marshall *(Science,* Vol. 227, pp. 729–30, Feb. 15, 1985).

P. 214 . . . knowledge is . . . guarded . . . An open question is the extent to which non-public procedures embodying some due process principles can improve the judging of classified information.

P. 215 . . . an experimental procedure . . . Reported in " 'Science court' would tackle knotty technological issues," by Leon Lindsay *(Christian Science Monitor,* p. 7, Mar. 23, 1983).

P. 215 . . . Roger Fisher and William Ury . . . See *Getting to Yes* (Boston: Houghton Mifflin Company, 1981).

P. 215 . . . Both sides . . . The procedures described here treat issues as two-sided, but this may seem too limited, because "issues," as commonly understood, often have many sides. In the energy debate, for example, gas, coal, nuclear, and solar power all have their advocates. Yet multisided issues contain many two-sided questions: Is the probability of a reactor meltdown low or high? Are the effects of coal burning on acid rain small or large? Will a solar collector cost little or much? Are gas reserves small or large? Multisided issues thus often resolve at their factual roots into numerical micro-questions.

Judicious scientists and engineers will seldom argue for high or low numbers as such; they will argue for the particular numbers they think most likely, or simply state evidence. But since holding a forum presupposes a dispute, advocates will be involved, and they will often wish to push far in one direction—nuclear advocates would like to prove that reactors are very cheap and safe; their opponents would like to prove that they are very expensive and deadly. Because numbers measuring cost and risk can only be larger or smaller, these micro-questions will tend to be two-sided.

CHAPTER 14: THE NETWORK OF KNOWLEDGE

P. 217 . . . Tohru Moto-oka . . . He is a professor at Tokyo University and the titular head of Japan's Fifth Generation Computer Project.

P. 220 . . . one system's structure . . . Their approach to hypertext, now in the demonstration stage, is called the Xanadu system. I have examined the proprietary data structures on which their system is based, and it is clear that powerful hypertext systems are indeed possible. For a less ambitious yet still quite powerful system, see "A Network-Based Approach to Text-Handling for the Online Scientific Community," a thesis by Randall H. Trigg (University of Maryland, Department of Computer Science, TR-1346, Nov. 1983).

P. 221 . . . Theodor Nelson's books . . . See *Computer Lib/Dream Machines* (self-published, distributed by The Distributors, South Bend, Ind., 1974), and *Literary Machines* (Swarthmore, Pa.: Ted Nelson, 1981). *Computer Lib* is an entertaining and idiosyncratic view of computers and their potential, including hypertext; a new edition is in preparation. *Literary Machines* focuses on hypertext.

P. 229 . . . *Time* magazine reports . . . on p. 76, June 13, 1983.

P. 230 . . . increasing the quantity of information available . . . A hypertext system might store the most commonly used information in the home, or in a local branch library. Compact disks of the sort used for audio recordings cost about three dollars to manufacture and can store as much text as about 500 books. See "Audio Analysis II: Read-only Optical Disks," by Christopher Fry (*Computer Music Journal*, Vol. 9, Summer 1985).

CHAPTER 15: WORLDS ENOUGH, AND TIME

P. 231 . . . bring abundance and long life to all who wish them . . .
But the limits to exponential growth ensure that universal, unconditional
abundance cannot last indefinitely. This raises questions regarding the distribution and ownership of space resources. Three basic approaches might
be considered:

One is a first-come, first-served approach, like the claiming of home-
steads or mining sites through use. This has roots in the Lockean principle
that ownership may be established by mixing one's labor with a previously
unowned resource. But this might allow a person with a suitable replicator
to turn it loose in space to rework—and thus claim—every unclaimed
object in the universe, as fast as it could be reached. This winner-take-all
approach has little moral justification, and would have unpleasant consequences.

A second extreme would be to distribute ownership of space resources
equally among all people, and to keep redistributing them to maintain
equality. This, too, would have unpleasant consequences. In the absence
of universal, stringent, compulsory limitations on childbearing, some
groups would continue to grow exponentially; evolutionary principles virtually guarantee this. In a surprisingly short time, the result of endless
redistribution would be to drag the standard of living of *every* human
being down to the minimum level that allows *any* group to reproduce.
This would mean hunger and poverty more extreme and universal than
that of any Third World country. If 99 percent of the human race volunta-
rily limited its birth rate, this would merely allow the remaining one percent to expand until it absorbed almost all the resources.

A third basic approach (which has many variations) takes a middle path:
it involves distributing ownership of the resources of space (genuine, permanent, transferable ownership) equally among all people—but doing so
only *once,* then letting people provide for their progeny (or others') from
their own vast share of the wealth of space. This will allow different
groups to pursue different futures, and it will reward the frugal rather than
the profligate. It can provide the foundation for a future of unlimited
diversity for the indefinite future, if active shields are used to protect
people from aggression and theft. No one has yet voiced a plausible alternative.

From a socialist perspective, this approach means equal riches for all.
From a libertarian perspective, it violates no one's property rights and
provides a basis for a future of liberty. In Thomas Schelling's terms, equal
division is a focal point solution in a coordination game (see *The Strategy of
Conflict,* by Thomas Schelling, Cambridge, Mass.: Harvard University

Press, 1960). What "equal division" actually means is a messy question best left to lawyers.

For this approach to work, agreement will be needed not just on a principle of division, but on a date. Space has been declared by treaty to be "the common heritage of all mankind," and we need to choose an Inheritance Day. Schelling's analysis suggests the importance, in a coordination game, of finding a *specific,* plausible proposal and of making it visible as soon as possible. Does a date suggest itself? A round-numbered space-related anniversary would seem appropriate, if it were not tied exclusively to the U.S. or U.S.S.R., or too soon, or too near a millennial date on the calendar. These constraints can be met; the most plausible candidate is perhaps April 12, 2011: the thirtieth anniversary of the flight of the world's first reusable spacecraft, the space shuttle, and the fiftieth anniversary of the flight of the first human into space, Yuri Gagarin.

If, before this date, someone finds and employs a means to raise human reproduction rates by a factor of ten or more, then Inheritance Day should immediately be made retroactive to April 12 of the preceding year, and the paperwork sorted out later.

P. 232 . . . to secure a stable, durable peace . . . Active shields can accomplish this reliably only through the use of redundancy and ample safety margins.

P. 234 . . . but nature seems uncooperative . . . For a discussion of the apparent impossibility of time machines in general relativity, see "Singularities and Casuality Violation," by Frank J. Tipler in *Annals of Physics,* Vol. 108, pp. 1–36, 1977. Tipler is open-minded; in 1974 he had argued the other side of the case.

P. 236 . . . patterns that resemble . . . "fractals" . . . Fractal patterns have similar parts on different scales—as, a twig may resemble a branch which in turn resembles a tree, or as gullies, streams, and rivers may all echo each other's forms. See *The Fractal Geometry of Nature,* by Benoit B. Mandelbrot (San Francisco: W. H. Freeman, 1982).

AFTERWORD

P. 240 . . . this was actually accomplished in 1988 . . . A good review of this and related work may be found in "Protein Design, a Minimalist Approach," by William F. DeGrado, Zelda R. Wasserman, and James D. Lear (*Science,* Vol. 243, pp. 622–28, 1989).

P. 240 . . . a Nobel prize was shared . . . Of particular interest are the Nobel lectures of the two currently active researchers. See "Supramolecular Chemistry—Scope and Perspectives: Molecules, Supermolecules, and Molecular Devices," by Jean-Marie Lehn (*Angewandte Chemie International Edition in English,* Vol. 27, pp. 89-112, 1988) and "The Design of Molecular Hosts,

Guests, and Their Complexes," by Donald J. Cram (*Science*, Vol. 240, pp. 760–67, 1988).

P. 240 . . . observed and modified individual molecules . . . See "Molecular Manipulation Using a Tunnelling Microscope," by J. S. Foster, J. E. Frommer, and P. C. Arnett (*Nature*, Vol. 331, pp. 324–26, 1988).

P. 240 . . . Computer-based tools . . . Software useful for computer-aided design of protein molecules has been described in "Computer-Aided Model-Building Strategies for Protein Design," by C. P. Pabo and E. G. Suchanek (*Biochemistry*, Vol. 25, pp. 5987–91, 1986), and in "Knowledge-based Protein Modelling and Design," by Tom Blundell et al. (*European Journal of Biochemistry*, Vol. 172, pp. 513–20, 1988); a program which reportedly yields excellent results in designing hydrophobic side-chain packings for protein core regions is described by Jay W. Ponder and Frederic M. Richards in "Tertiary Templates for Proteins" (*Journal of Molecular Biology*, Vol. 193, pp. 775–91, 1987). The latter authors have also done work in molecular modeling (an enormous and active field); see "An Efficient Newton-like Method for Molecular Mechanics Energy Minimization of Large Molecules" (*Journal of Computational Chemistry*, Vol. 8, pp. 1016–24, 1987). Computational techniques derived from molecular mechanics have been used to model quantum effects on molecular motion (as distinct from quantum-mechanical modeling of electrons and bonds); see "Quantum Simulation of Ferrocytochrome c," by Chong Zheng et al. (*Nature*, Vol. 334, pp. 726–28, 1988).

P. 241 . . . A recent summary . . . K. Eric Drexler, "Machines of Inner Space," in *1990 Yearbook of Science and the Future*, edited by D. Calhoun, pp. 160–77 (Chicago: Encyclopaedia Britannica, 1989).

P. 241 . . . A variety of technical papers . . . These include the following papers (which will be collected and rewritten as parts of my forthcoming technical book): "Nanomachinery: Atomically Precise Gears and Bearings," in the proceedings of the IEEE Micro Robots and Teleoperators Workshop (Hyannis, Massachusetts: IEEE, 1987); "Exploring Future Technologies," in *The Reality Club*, edited by J. Brockman, pp. 129–50 (New York: Lynx Books, 1988); "Biological and Nanomechanical Systems: Contrasts in Evolutionary Capacity," in *Artificial Life*, edited by C. G. Langton, pp. 501–19 (Reading, Massachusetts: Addison-Wesley, 1989); and "Rod Logic and Thermal Noise in the Mechanical Nanocomputer," in *Molecular Electronic Devices III*, edited by F. L. Carter (Amsterdam: Elsevier Science Publishers B.V., in press). For information on the availability of technical papers, please contact the Foresight Institute at the address given in the Afterword.

GLOSSARY

This glossary contains terms used in describing matters related to advanced technology. It was compiled by the MIT Nanotechnology Study Group, with special help from David Darrow of Indiana University.

ACTIVE SHIELD: A defensive system with built-in constraints to limit or prevent its offensive use.

AMINO ACIDS: Organic molecules that are the building blocks of proteins. There are some two hundred known amino acids, of which twenty are used extensively in living organisms.

ANTIOXIDANTS: Chemicals that protect against oxidation, which causes rancidity in fats and damage to DNA.

ARTIFICIAL INTELLIGENCE (AI): A field of research that aims to understand and build intelligent machines; this term may also refer to an intelligent machine itself.

ASSEMBLER: A molecular machine that can be programmed to build virtually any molecular structure or device from simpler chemical building blocks. Analogous to a computer-driven machine shop. (See Replicator.)

ATOM: The smallest particle of a chemical element (about three ten-billionths of a meter in diameter). Atoms are the building blocks of molecules and solid objects; they consist of a cloud of electrons surrounding a dense nucleus a hundred thousand times smaller than the atom itself. Nanomachines will work with atoms, not nuclei.

AUTOMATED ENGINEERING: The use of computers to perform engineering design, ultimately generating detailed designs from broad specifications with little or no human help. Automated engineering is a specialized form of artificial intelligence.

BACTERIA: One-celled living organisms, typically about one micron in diameter. Bacteria are among the oldest, simplest, and smallest types of cells.

BIOCHAUVINISM: The prejudice that biological systems have an intrinsic superiority that will always give them a monopoly on self-reproduction and intelligence.

BIOSTASIS: A condition in which an organism's cell and tissue structures are preserved, allowing later restoration by cell repair machines.

BULK TECHNOLOGY: Technology based on the manipulation of atoms and molecules in bulk, rather than individually; most present technology falls in this category.

CAPILLARIES: Microscopic blood vessels that carry oxygenated blood to tissues.

CELL: A membrane-bound unit, typically microns in diameter. All plants and animals are made up of one or more cells (trillions, in the case of human beings). In general, each cell of a multicellular organism contains a nucleus holding all of the genetic information of the organism.

CELL REPAIR MACHINE: A system including nanocomputers and molecular-scale sensors and tools, programmed to repair damage to cells and tissues.

CHIP: See Integrated circuit.

CROSS-LINKING: A process forming chemical bonds between two separate molecular chains.

CRYOBIOLOGY: The science of biology at low temperatures; research in cryobiology has made possible the freezing and storing of sperm and blood for later use.

CRYSTAL LATTICE: The regular three-dimensional pattern of atoms in a crystal.

DESIGN AHEAD: The use of known principles of science and engineering to design systems that can only be built with tools not yet available; this permits faster exploitation of the abilities of new tools.

DESIGN DIVERSITY: A form of redundancy in which components of different design serve the same purpose; this can enable systems to function properly despite design flaws.

DISASSEMBLER: A system of nanomachines able to take an object apart a few atoms at a time, while recording its structure at the molecular level.

DISSOLUTION: Deterioration in an organism such that its original structure cannot be determined from its current state.

DNA (DEOXYRIBONUCLEIC ACID): DNA molecules are long chains consisting of four kinds of nucleotides; the order of these nucleotides encodes the information needed to construct protein molecules. These in turn make up much of the molecular machinery of the cell. DNA is the genetic material of cells. (See also RNA.)

ENGINEERING: The use of scientific knowledge and trial-and-error to design systems. (See Science.)

ENTROPY: A measure of the disorder of a physical system.

ENZYME: A protein that acts as a catalyst in a biochemical reaction.

EURISKO: A computer program developed by Professor Douglas Lenat which is able to apply heuristic rules for performing various tasks, including the invention of new heuristic rules.

EVOLUTION: A process in which a population of self-replicating entities undergoes variation, with successful variants spreading and becoming the basis for further variation.

EXPONENTIAL GROWTH: Growth that proceeds in a manner characterized by periodic doublings.

FACT FORUM: A procedure for seeking facts through a structured, arbitrated debate between experts.

FREE RADICAL: A molecule containing an unpaired electron, typically highly unstable and reactive. Free radicals can damage the molecular machinery of biological systems, leading to cross-linking and mutation.

HEISENBERG UNCERTAINTY PRINCIPLE: A quantum-mechanical principle with the consequence that the position and momentum of an object cannot be precisely determined. The Heisenberg principle helps determine the size of electron clouds, and hence the size of atoms.

HEURISTICS: Rules of thumb used to guide one in the direction of probable solutions to a problem.

HYPERTEXT: A computer-based system for linking text and other information with cross-references, making access fast and criticisms easy to publish and find.

INTEGRATED CIRCUIT (IC): An electronic circuit consisting of many interconnected devices on one piece of semiconductor, typically 1 to 10 millimeters on a side. ICs are the major building blocks of today's computers.

ION: An atom with more or fewer electrons than those needed to cancel the electronic charge of the nucleus. An ion is an atom with a net electric charge.

KEVLAR (TM): A synthetic fiber made by E. I. du Pont de Nemours & Co., Inc. Stronger than most steels, Kevlar is among the strongest commercially available materials and is used in aerospace construction, bulletproof vests, and other applications requiring a high strength-to-weight ratio.

LIGHTSAIL: A spacecraft propulsion system that gains thrust from the pressure of light striking a thin metal film.

LIMITED ASSEMBLER: An assembler with built-in limits that constrain its use (for example, to make hazardous uses difficult or impossible, or to build just one thing).

MEME: An idea that, like a gene, can replicate and evolve. Examples of memes (and meme systems) include political theories, proselytizing religions, and the idea of memes itself.

MOLECULAR TECHNOLOGY: See Nanotechnology.

MOLECULE: The smallest particle of a chemical substance; typically a group of atoms held together in a particular pattern by chemical bonds.

MUTATION: An inheritable modification in a genetic molecule, such as DNA. Mutations may be good, bad, or neutral in their effects on an organism; competition weeds out the bad, leaving the good and the neutral.

NANO-: A prefix meaning ten to the minus ninth power, or one billionth.

NANOCOMPUTER: A computer made from components (mechanical, electronic, or otherwise) on a nanometer scale.

NANOTECHNOLOGY: Technology based on the manipulation of individual atoms and molecules to build structures to complex, atomic specifications.

NEURAL SIMULATION: Imitating the functions of a neural system—such as the brain—by simulating the function of each cell.

NEURON: A nerve cell, such as those found in the brain.

NUCLEOTIDE: A small molecule composed of three parts: a nitrogen base (a purine or pyrimidine), a sugar (ribose or deoxyribose), and phosphate. Nucleotides serve as the building blocks of nucleic acids (DNA and RNA).

NUCLEUS: In biology, a structure in advanced cells that contains the chromosomes and apparatus to transcribe DNA into RNA. In physics, the small, dense core of an atom.

ORGANIC MOLECULE: A molecule containing carbon; the complex molecules in living systems are all organic molecules in this sense.

POLYMER: A molecule made up of smaller units bonded to form a chain.

POSITIVE SUM: A term used to describe a situation in which one or more entities can gain without other entities suffering an equal loss—for example, a growing economy. (See Zero sum.)

REDUNDANCY: The use of more components than are needed to perform a function; this can enable a system to operate properly despite failed components.

REPLICATOR: In discussions of evolution, a replicator is an entity (such as a gene, a meme, or the contents of a computer memory disk) which can get itself copied, including any changes it may have undergone. In a broader sense, a replicator is a system which can make a copy of itself, not necessarily copying any changes it may have undergone. A rabbit's genes are replicators in the first sense (a change in a gene can be inherited); the rabbit itself is a replicator only in the second sense (a notch made in its ear can't be inherited).

RESTRICTION ENZYME: An enzyme that cuts DNA at a specific site, allowing biologists to insert or delete genetic material.

RIBONUCLEASE: An enzyme that cuts RNA molecules into smaller pieces.

RIBOSOME: A molecular machine, found in all cells, which builds protein molecules according to instructions read from RNA molecules. Ribosomes are complex structures built of protein and RNA molecules.

RNA: Ribonucleic acid; a molecule similar to DNA. In cells, the information in DNA is transcribed to RNA, which in turn is "read" to direct protein construction. Some viruses use RNA as their genetic material.

SCIENCE: The process of developing a systematized knowledge of the world through the variation and testing of hypotheses. (See Engineering.)

SCIENCE COURT: A name (originally applied by the media) for a government-conducted fact forum.

SEALED ASSEMBLER LABORATORY: A work space, containing assemblers, encapsulated in a way that allows information to flow in and out but does not allow the escape of assemblers or their products.

SYNAPSE: A structure that transmits signals from a neuron to an adjacent neuron (or other cell).

VIRUS: A small replicator consisting of little but a package of DNA or RNA which, when injected into a host cell, can direct the cell's molecular machinery to make more viruses.

ZERO SUM: A term used to describe a situation in which one entity can gain only if other entities suffer an equal loss; for example, a private poker game. (See Positive sum.)

Index

ABOUT THE AUTHOR

K. Eric Drexler is a researcher concerned with emerging technologies. Since 1977, his work has increasingly focused on nanotechnology, a development which many now view as a basic manufacturing technology of the twenty-first century. A graduate of the Massachusetts Institute of Technology, he is currently a visiting scholar at Stanford University (where he taught the first university course on nanotechnology) and president of the Foresight Institute, a nonprofit organization with the mission of "preparing for future technologies." He lectures widely on nanotechnology and its consequences.